SECOND EDITION

INTRODUCTION TO THE SPACE ENVIRONMENT

by
Thomas F. Tascione

KRIEGER PUBLISHING COMPANY
MALABAR, FLORIDA
2010

Original Edition 1988
Second Edition 1994
Second Edition Reprint w/new material 2010
ISBN 10: 0-89464-071-2
ISBN 13: 978-0-89464-071-1

Printed and Published by
KRIEGER PUBLISHING COMPANY
KRIEGER DRIVE
MALABAR, FLORIDA 32950

FROM A DECLARATION OF PRINCIPLES JOINTLY ADOPTED BY A COMMITTEE OF THE AMERICAN BAR ASSOCIATION AND A COMMITTEE OF PUBLISHERS:
This publication is designed to provide accurate and authoritative information in regard to the subject matter covered. It is sold with the understanding that the publisher is not engaged in rendering legal, accounting, or other profes-sional service. If legal advice or other expert assistance is required, the services of a competent professional person should be sought.

Library of Congress Cataloging-In-Publication Data

Tascione, Thomas F.
      Introduction to the space environment / Thomas F. Tascione.—2nd
    ed.
            p.   cm.
      Includes index.
      ISBN 0-89464-044-5 (alk. paper)
      1. Atmosphere, Upper. 2. Sun. 3. Ionosphere. 4. Space plasmas.
I. Title.
      QC879.T37   1994
      551.5'14—dc20
                                                                93-36569
                                                                CIP

10   9   8   7   6   5   4   3   2

To my family

Series Editor
Edwin F. Strother, Ph.D.

Contents

Preface to the Second Edition

Since the publication of the first edition, scientific discovery in space science has continued at an extraordinary rate. These discoveries have led to new levels of understanding of the basic physical processes controlling the space environment. The purpose of the second edition is to capture these ideas and present them in a clear and concise fashion for the novice. The reader is expected to have a solid background in introductory physics and therefore this edition is most useful as a textbook for senior year college, or first-year graduate, students.

At the request of colleagues and students, I have revised and updated the original text and added a considerable amount of new material. In many cases, I had to simplify the mathematical treatment to keep the book at the level of the intended reader. However, in each of these cases I tried to avoid oversimplifying ideas to the point where they lose their physical meaning. I also tried to avoid judging between conflicting theories for which there is no clear-cut observational evidence to help us choose between one theory or the other. For these cases, I present opposing points of view and indicate how further observational evidence might be able to resolve the differences of opinion.

The book still contains ten chapters, but new topics have been added to all but one chapter.

Chapter 1 remains a review of those basic principles of plasma physics that form the foundation of topics developed later in the text. Added topics include a physical explanation of "frozen-in" magnetic flux and a discussion of the secondary ionization processes occurring in the auroral regions.

The focus of Chapter 2 is the basic physical processes controlling the solar atmosphere. Additions include more discussion on the properties of coronal holes and coronal mass ejections, the characteristics and frequency of occurrence of solar proton events, a description of the Zurich and McIntosh sunspot classification systems, and the use of the chromospheric neutral lines as a predictor of potential solar activity.

Chapter 3 describes the physics controlling the formation and dynamics of the solar wind and interplanetary magnetic field. New sections include *Solar Wind Power Generation* and *Libration Point Measurements*. The section *Solar Wind Power Generation* develops the physical principles behind the "epsilon" factor which can provide a measure of the "available" electrical power in the solar wind to drive the magnetosphere. The *Libration Point Measurements* section discusses the data collection capabilities of a satellite at an L1 parking orbit and how this data can be used to improve our understanding of how the solar wind drives magnetospheric disturbances.

Chapter 4 discusses the structure and origin of the geomagnetic field. Additions to Chapter 4 include a discussion on the physics behind the equatorial electroject and how the adiabatic invariants, derived in Chapter 1, apply to particles trapped in the radiation belts.

Chapter 5 details the physics of the solar wind interaction with the geomagnetic field and the resulting dynamic region called the magnetosphere. The second edition includes four new sections and subjections: *Collisionless Shocks: Wave Particle Interactions*; *Ring Current*; *Field-Aligned Currents*; and *Observations: Growth Phase Versus Directly Driven Model*. The section *Collisionless Shocks: Wave Particle Interactions* describes the physical characteristics of ion-acoustic waves, Landau damping, and the hydromagnetic Alfven waves as applied to the Earth's magnetosphere. The subsections *Ring Current* and *Field-Aligned Currents* describe the physics behind these magnetospheric current systems. *Observations: Growth Phase Versus Directly Driven Model* describes some of the available experimental data used to support the two competing models describing auroral substorms. In addition, the sections *Magnetosheath* and *Plasmasphere* have been extensively revised and updated.

Chapter 6 describes the physical processes controlling the stratosphere, mesosphere, and thermosphere. Additions include more details on variations in the ozone layer including the "ozone hole," and a more thorough discussion of the physics controlling the thermospheric temperature and density variability as a function of the solar cycle and geomagnetic activity.

Chapter 7 develops the theory of how solar radiation forms an ionosphere. This chapter remains unchanged from the first edition.

Chapter 8 describes the deviations from the classic solar driven ionosphere due to ionospheric convection and geomagnetic activity. The second edition includes more discussion of the following topics: foF2 variation with changes in geomagnetic activity, the $\vec{E} \times \vec{B}$ drift instability and low-latitude scintillation, high-latitude convection patterns, inverted-V electron beams, polar rain, and the theta aurora.

Chapter 9 and 10 apply the theory developed in the earlier chapters. In Chapter 9, we discover how the ionosphere affects radiowave propagation. Additions to Chapter 9 include more information on the latitudinal, seasonal, and solar cycle variation of the scintillation phenomena which can degrade ground-to-satellite or satellite-to-ground communications. Chapter 10 describes how the space environment can impact spacecraft operations. Two new sections are *Deep Dielectric Charging* and *Space Shuttle Glow*. In addition, the Chapter 10 radiation dose tables have been updated with more recent observational data.

Chapter 11, added in 2009, defines the space weather nowcast and forecast requirements for the two segments of the commercial space weather community: consumers and service providers. The consumer segment includes system builders, system operators, and system users, each with its unique need for space weather services.

As in the first edition, the new edition does not reference individual authors within the text itself. This is done as a personal preference in the belief it makes the text easier to read. It is not an attempt on my part to take credit for ideas or discoveries which are not my own. A complete list of references is given at the end of each chapter.

Preface to the First Edition

Our understanding of the near Earth space environment has grown rapidly through the decade of the 1970s because of the detailed measurements of many sophisticated, in situ satellite platforms. The intent of the book is to summarize the complicated and sometimes conflicting theories which have evolved from these measurements in a clear and readable fashion for the novice. However, the book does assume the reader has a firm grasp of introductory physics and is therefore directed toward the senior year college or first-year graduate student.

The book is not intended to be on the leading edge of space science by introducing new and revolutionary ideas. Instead, the book summarizes the popular ideas at the time of writing. Ideas evolve in time, however, and years from now our knowledge may be very different from the theories contained within this text. Such evolution should be encouraged and applauded because space physics is a new and vibrant science.

The book is organized into ten chapters. Chapter 1 is a review of plasma physics and I apologize to any plasma physicist for the omissions and sometimes terse explanations used. In my defense, this is not a plasma physics book, but rather a book which shows how plasma physics is applied to the space environment. It is assumed that the concepts of electric and magnetic fields are not completely foreign to the reader. The chapter is a review of terminology and concepts which will be used later in the text.

Chapter 2 reviews solar physics and again there are omissions and short summaries of ideas and theories. This chapter is also intended as a review of important concepts which will be used later in the text. A lengthy reference list at the end of the chapter will direct the interested reader to detailed expositions.

Chapter 3 discusses the formation and dynamics of the solar wind and interplanetary magnetic field. Although some of the models developed are oversimplified, they do bring out all the important characteristics of the complicated plasma dynamics.

Chapter 4 reviews the structure and origin of the geomagnetic field. Also included is a discussion of geomagnetic disturbances.

Chapter 5 deals with the interaction of the solar wind with the geomagnetic field. The resulting dynamical region, called the magnetosphere, has a profound effect on any near Earth satellite operations.

Chapter 6 reviews the structure of the neutral atmosphere which provides the basic constituents for the ionosphere. It is shown that the neutral winds are very important to dynamics of the lower ionosphere.

Chapter 7 develops the theory of ionospheric formation and applies this theory to various ionospheric layers. The dynamical nature of the environment is shown to be a crucial feature in understanding the overall ionospheric structure.

Chapter 8 discusses how the ionosphere and magnetosphere are coupled. It is shown that the ionosphere is quite different with latitude and this difference is a result of the overall magnetosphere-ionosphere system.

The last two chapters apply the ideas developed in earlier chapters. Chapter 9 looks at ground based systems, especially long-range communications, while Chapter 10 examines how the space environment affects both manned and unmanned space systems.

Unlike a journal article, individual authors are not referenced within the text itself. This was done to make the material easier to read and is not an attempt on my part to take credit for ideas which are not my own. A complete list of references is given at the end of each chapter.

Finally, I would like to acknowledge the tireless and unselfish work of Peggy Raymond, Norma Cox, and Cindy Gearhart who deciphered my handwriting and put the text in a readable format. Ladies, I thank you from the bottom of my heart.

Preface to the First Edition

Chapter 1

Plasma Physics

1.0 Introduction

A plasma is usually defined as an electrically neutral, ionized gas. A gas can be both ionized and electrically neutral at the same time, provided there are as many free electrons in the gas as there are net positive charges on the positive gas ions. Thus, a plasma consists of a homogeneous mixture of electrons and positive ions, surging and swirling in electromagnetic bondage. The collective motions of charged particles generate magnetic fields that, in turn, govern the charged particle motions themselves. Individual charged particles follow magnetically bent paths as they move between electrostatic collisions.

The Sun itself is a huge, dense swirl of plasma, radiating away its excess thermonuclear energy as streams of photons and charged particles. The charged particles then form a tenuous plasma that sweeps through interplanetary space. Thus, an understanding of the very environment of space itself hinges on gaining an understanding of plasmas. Furthermore, plasmas are involved in a large part of our everyday lives. When you turn the key to start your car, the flash of electricity across the spark plug gap forms a plasma. The stroke of lightning across the summer sky forms a plasma too, just as does the flare of a match.

In this chapter we will discuss some of the applicable physical laws and concepts, some of the motions within a plasma, the behavior of a magnetic field in a plasma, and the behavior of a plasma on encountering a magnetic field. The later parts of the chapter introduce the ways that electromagnetic radiation interacts with plasmas.

1.1 Maxwell's Equations

Maxwell's equations are actually a collection of four very powerful laws—Gauss's law, Gauss's law for magnetism, Ampere's law, and Faraday's law. These laws are sometimes stated in terms that are difficult to decipher, but the concepts themselves are easy to understand. Gauss's law is a good example of this.

Gauss's law essentially states that the integral (over an entire Gaussian, or hypothetical, closed surface) of the product of an incremental area of the surface and the electric field vector normal to that surface is proportional to the net electric charge enclosed by the surface. Symbolically

$$\epsilon_o \int \vec{E} \cdot \vec{ds} = q \qquad (1.1)$$

where $\vec{E}$ is the electric field, $\vec{ds}$ is the element of surface area (whose direction is the outward-drawn normal to the surface), ϵ_o is the permittivity of free space and q is the net electrical charge enclosed. Physically, the easiest way to visualize Gauss's law is to imagine a fluid flowing away from a source. If we enclose the source in an open mesh bag, it is obvious that all the fluid emitted from the source must flow out of the bag, regardless of the bag's shape. If one could measure the amount of fluid passing through the bag, then one can compute the intensity of the source. Returning to equation 1.1, the electric field is our "fluid" which emanates from the source of free electrical charges. The problem is complicated by the fact that electric fields can pass *into*, as well as *out of*, our imaginary surface.

Gauss's law for magnetism is simply the application of the above ideas to magnetic lines of force. However, there are no known isolated sources of magnetic field lines of a given direction (i.e., monopoles). Experience indicates that single magnetic poles do not exist; for every north magnetic pole there is an accompanying south magnetic pole. Notice that we used the word experience to indicate how nature likes to organize magnetic poles because, theoretically, there is no reason to assume monopoles do not exist. However, given that no verifiable monopoles have been reported to date, Gauss's law for magnetism is written as

$$\int \vec{B} \cdot \vec{ds} = 0. \qquad (1.2)$$

This equation implies that the magnetic lines of force are continuous and, therefore, close on themselves. The number of lines of force leaving our imaginary surface must be balanced by an equal number of lines entering into the surface.

The third law, Ampere's law, states that an electric current produces a magnetic field which encircles the current; the stronger the current, the stronger the magnetic field. That is,

$$\int \vec{B} \cdot \vec{dl} = \mu_o \int \vec{J} \cdot \vec{dA} \qquad (1.3)$$

where $\vec{B}$ is the magnetic field, $\vec{dl}$ is a differential line element along an arbitrary hypothetical closed loop drawn in a region where the magnetic fields and current distributions exist, $\overrightarrow{J}$ is the current density, $d\overrightarrow{A}$ is an element of area through which the current passes, and μ_o is the magnetic permeability of free space. Maxwell found that Ampere's law was incomplete. Experiments showed that a time rate of change in the electric flux also produced a magnetic field, and this additional "current like" phenomena is called the displacement current. The electric flux can be written as

$$\Phi_E = \int \overrightarrow{E} \cdot d\overrightarrow{A} \qquad (1.4)$$

and the Maxwell generalized equation 1.3 as

$$\int \overrightarrow{B} \cdot d\overrightarrow{l} = \mu_o \epsilon_o \frac{d\Phi_E}{dt} + \mu_o \int \overrightarrow{J} \cdot d\overrightarrow{A}. \qquad (1.5)$$

The last of Maxwell's laws is Faraday's law of magnetic induction which states that a time rate of change of magnetic flux will induce its own electric field. Mathematically,

$$\oint \overrightarrow{E} \cdot d\overrightarrow{l} = -\frac{d\Phi_B}{dt} = -\frac{d}{dt}\int \overrightarrow{B} \cdot d\overrightarrow{A} \qquad (1.6)$$

where the first integral is measured around the enclosed path within which the changing magnetic flux is measured.

Whether we know it or not, we are all familiar with the results of Faraday's law because it is the basic principle behind electric generators and transformers.

1.2 Ohm's Law

The last two laws of Maxwell's equations deal with electric currents (either directly or indirectly), and therefore, we should spend some time discussing the impedance to this current flow. Ohm's law parameterizes the impedance effects as a linear relation between the electric current and the sources of potential difference which make the current flow. Possible sources of potential differences are electric fields, magnetic fields, pressure gradients, and gravitational forces. It is customary to consider the proportionality constant as a conductance (the inverse of resistance). Furthermore, the generalized vectorial form of Ohm's law is almost invariably expressed in terms of a unit volume and thus the electric current I becomes the electric current density $\overrightarrow{J}$. For example, if the *only* source of potential difference is an electric field then Ohm's law becomes

$$\overrightarrow{J} = \overrightarrow{\sigma} \cdot \overrightarrow{E}. \qquad (1.7)$$

where $\overrightarrow{\sigma}$ is the conductivity tensor. As we will soon see, the mobility (or conductivity) of charged particles depends on the magnetic field orientation within a plasma. That is, the mobility across magnetic field lines is much different than the mobility along field lines. We refer the interested reader to Spitzer's book (see references) for the more general form of Ohm's law.

The usual mathematical description uses a coordinate system drawn so that the z axis is parallel to $\overrightarrow{B}$ and the direction of the electric field is broken into components parallel or perpendicular to $\overrightarrow{B}$. In such a coordinate system the conductivity consists of three components:

$$\sigma_P, \sigma_H \text{ and } \sigma_\parallel$$

which are called the Pedersen, Hall and parallel conductivities respectively, and the conductivity tensor, $\overrightarrow{\sigma}$, becomes

$$\overrightarrow{\sigma} = \begin{pmatrix} \sigma_P & \sigma_H & 0 \\ -\sigma_H & \sigma_P & 0 \\ 0 & 0 & \sigma_\parallel \end{pmatrix}.$$

The Pederson conductivity is along a path parallel to $\overrightarrow{E}$ and perpendicular to $\overrightarrow{B}$. The Hall conductivity is along a path perpendicular to both $\overrightarrow{E}$ and $\overrightarrow{B}$, and the parallel conductivity is along the direction of $\overrightarrow{B}$. In the limit that ion-electron collisions are unimportant we have

$$\sigma_P = \left[\frac{\nu_{en}\omega_{ce}}{\nu_{en}^2 + \omega_{ce}^2} + \frac{\nu_{in}\omega_{ci}}{\nu_{in}^2 + \omega_{ci}^2} \right] \frac{e N_e}{B} \qquad (1.8)$$

$$\sigma_H = \left[\frac{\omega_{ci}^2}{\nu_{in}^2 + \omega_{ci}^2} - \frac{\omega_{ce}^2}{\nu_{en}^2 + \omega_{ce}^2} \right] \frac{e N_e}{B} \qquad (1.9)$$

$$\sigma_\parallel = \left[\frac{1}{M_e \nu_{en}} + \frac{1}{M_i \nu_{in}} \right] e^2 N_e \qquad (1.10)$$

where ν_{en} and ν_{in} are the electron-neutral and ion-neutral collision frequencies respectively, ω_{ce} and ω_{ci} are the electron and ion cyclotron frequencies (see section 1.6), M_e and M_i are the electron and ion masses, and N_e is the electron number density.

An important property of a highly conducting plasma is that the magnetic field becomes "frozen-in" the plasma (i.e., the magnetic field is carried along within the plasma virtually unchanged). In particular, we would like to develop the form of the frozen flux equation as seen for an observer on Earth watching the solar wind streaming past. In the vicinity of the Earth, observational evidence indicates that the interplanetary electric field is nearly always perpendicular to the velocity vector; usually there is not an electric field parallel to $\overrightarrow{v}$ which would increase the solar wind speed. As seen from the Earth, a perpendicular interplanetary electric field becomes

$$\overrightarrow{E}_w = \frac{\overrightarrow{E} + \overrightarrow{v} \times \overrightarrow{B}}{\sqrt{1 - (v/c)^2}} \qquad (1.11)$$

where $\overrightarrow{E}_w$ is the electric field seen by a particle in the solar wind, $\overrightarrow{E}$ is the solar wind electric field at Earth, $\overrightarrow{v}$ is the solar wind velocity, $\overrightarrow{B}$ is the interplanetary magnetic field, and v/c is the ratio of the solar wind velocity to the speed of light. For normal solar wind conditions

$$\frac{v}{c} \ll 1 \qquad (1.12)$$

and equation 1.7 for the electric current density becomes

$$\overrightarrow{J} = \overrightarrow{\sigma} \cdot (\overrightarrow{E} + \overrightarrow{v} \times \overrightarrow{B}). \qquad (1.13)$$

Combining equation 1.11 with equation 1.6 and applying Stokes' theorem gives

$$\frac{d\Phi_B}{dt} = -\int_A \nabla \times (\vec{E} + \vec{v} \times \vec{B}) \cdot d\vec{A} \quad (1.14)$$

where $d\vec{A}$ is the unit area with unit vector normal to and directed outward from the surface of integration. By definition, frozen-in flux conditions imply the magnetic flux, Φ_B, must remain constant with time as seen in the reference frame of the moving particle. Therefore, from equation 1.14, we see that frozen-in flux implies

$$\nabla \times (\vec{E} + \vec{v} \times \vec{B}) = 0. \quad (1.15)$$

The next step is to determine whether equation 1.15 can be simplified to

$$\vec{E} + \vec{v} \times \vec{B} = 0. \quad (1.16)$$

It is left as an exercise for the student (see Problem 1-5) to show that equation 1.16 is a valid description of frozen-in flux for the solar wind. Proof involves showing that the diffusion time for the interplanetary magnetic field is long compared to the time it takes for solar wind conditions to change.

1.3 Equation of Continuity

In general, the time rate of change of charged particles in any given unit volume is proportional to the net flow (in and out) of the volume plus creation and loss (ionization and recombination) of charged particles. That is,

$$\frac{\partial \rho}{\partial t} + -\nabla \cdot \rho \vec{v} + Q - L \quad (1.17)$$

where ρ is the mass density, $\vec{v}$ is the flow velocity, Q is the production rate, and L is the loss rate. The plasma mass density is usually written as

$$\rho = N_i M_i + N_e M_e \simeq N_e M_i \quad (1.18)$$

where we have assumed quasi-neutrality ($N_e \sim N_i$). Thus far we have assumed only one ion species, but we can easily generalize to any number of ion species by simply summing over the ion types in equation 1.18.

1.4 Hydrodynamic Equation

As with the continuity equation, it is useful in many cases to use the fluid-like characteristics of the plasma. The most correct formulation to use for particles is the Boltzmann transport equation, but due to its complexity, a number of simplifying assumptions are usually made to reduce it to a more tractable form. One such simpler form is the magneto-hydrodynamic equation of motion which is usually written as

$$\rho \frac{\partial \vec{v}}{\partial t} + \rho \vec{v} \cdot \nabla \vec{v} = -\nabla p + \rho \vec{g} + \vec{J} \times \vec{B} \quad (1.19)$$

where ρ is the plasma mass density, $\vec{v}$ is the bulk flow velocity, p is the gas pressure, $\vec{g}$ is the acceleration of gravity, $\vec{J}$ is the current density, and $\vec{B}$ is the magnetic field vector.

This equation and the other preceding ones must be solved simultaneously to yield meaningful results. Unfortunately it is extremely difficult to obtain solutions for even the simplest plasma cases. Furthermore, to succeed in obtaining any solution at all for some cases, it is necessary to make assumptions that eliminate whole classes of solutions. The hope, of course, is that the correct solution is not accidentally thrown away. It's rather like throwing away most of the haystack to make it easier to find the needle. In light of this, it is little wonder that plasma theory has been used more to explain what is observed, instead of to predict what will be observed.

1.5 Saha's Equation

Not all plasmas are fully ionized, and thus a portion of the volume occupied by a plasma may be occupied by a neutral gas. It is therefore of some interest to be able to determine what portion of the homogeneous mixture is ionized. One way to do this is by use of Saha's equation.

Saha derived the following equation for a homogeneous, thermal plasma (a random motion plasma in which the particle energies follow a Maxwell-Boltzmann distribution):

$$\frac{N_{q+1} N_e}{N_q} = \frac{(2\pi mkT)^{3/2}}{h^3} \frac{2\widetilde{\omega}_{q+1}}{\widetilde{\omega}_q} \exp\left(-\frac{X_q}{kT}\right) \quad (1.20)$$

where N_q is the density of a ion species in the qth state of ionization (i.e., 0^+), N_{q+1} is the density in the next higher state (i.e., 0^{++}), X_q is the ionization potential for state q, $\widetilde{\omega}$ is the statistical weight for a given level, k is Boltzmann's constant, and h is Planck's constant.

In interplanetary space, the highly ionized plasma is predominately hydrogen. Unfortunately, equation 1.20 is of little use because hydrogen has only one ionization state. However, equation 1.20 is *very useful* in ionospheric work where heavy atoms (e.g., oxygen) and molecules (e.g., O_2, CO_2, NO, etc.) predominate. Returning to interplanetary space, the simplest approach is to use a variation of the Saha equation. We can define the degree of ionization as the product of the positive ion and electron number densities, divided by the neutral atom number density. This in turn can be used to find the ionization ratio, r, which is the degree of ionization divided by the total number of atoms (both ionized and neutral) in a unit volume. That is,

$$r = \frac{N_i N_e}{N_o N} \quad (1.21)$$

where N_i is the ion density, N_e is the electron density, N_o is the neutral atom density and N is the total atom density. In interplanetary space it is reasonable to assume that for each free proton there is a free electron ($N_e \approx N_i$), and, since there are so few neutral atoms per unit volume, the total number density is approximately equal to the ion number density ($N = N_i + N_o \approx N_i$). Therefore, the ionization ratio becomes

$$r = \frac{N_i}{N_o} \qquad (1.22)$$

which is an enormous number for interplanetary space where $N_i \gg N_o$.

1.6 Single Particle Motion

In contrast to the bulk flow properties of plasmas discussed in earlier sections, we will now turn our attention to single particle motion of charged particles making up the plasma. In a sense, we will now examine the microscopic properties of the plasma, while our earlier discussion dealt with the macroscopic properties of the plasma.

With that in mind, we shall first consider the motion of a positively charged particle moving in the plane of this page. If there is also a magnetic field present which is directed *down into this page*, it can be seen by use of the right-hand rule that the positive charge will circle counterclockwise, and an electron would circle clockwise.

The radius of motion and the period of revolution are derived from the following argument. The force acting on a charged particle moving through a magnetic field is called the Lorentz force and is written as

$$\vec{F} = q \vec{v} \times \vec{B} \qquad (1.23)$$

where q is the charge intensity, $\vec{v}$ is the charged particle's velocity, and $\vec{B}$ is the magnetic field strength. The particle's path about B is a circle (see problem 1-1) in that the centripetal force is equal to the Lorentz force, and

$$-\frac{mv^2}{r}\hat{r} = q \vec{v} \times \vec{B} \qquad (1.24)$$

where r is the radius of gyration, m is the charged particle's mass, and $\hat{r}$ is the unit vector in the radial direction. If the particle's motion is at right angles to the magnetic field, equation 1.24 reduces to

$$\frac{mv^2}{r} = qvB. \qquad (1.25)$$

Solving for r yields the radius of gyration (gyroradius or cyclotron radius) which is denoted as r_c as follows:

$$r_c = \frac{mv}{qB}. \qquad (1.26)$$

The cyclotron frequency is simply

$$\nu = \frac{v}{2\pi r} = \frac{qB}{2\pi m} \qquad (1.27)$$

or in terms of angular frequency

$$\omega_c = 2\pi\nu = \frac{qB}{m}. \qquad (1.28)$$

Equation 1.28 is usually called the gyrofrequency for particle of charge q and mass m.

Since current I is charge per unit time, we can write the equivalent current for a gyrating charged particle as

$$I_e = q\left(\frac{1}{T}\right) = qf_c = \frac{q\omega_c}{2\pi}. \qquad (1.29)$$

Therefore, using equation 1.28 to eliminate ω_c yields

$$I_e = \frac{q\omega_c}{2\pi} = \frac{q^2B}{2\pi m}. \qquad (1.30)$$

For computational purposes the Earth's magnetic field can be considered to be on the order of 0.1 gauss or 10^{-5} webers/m^2 (10,000 gammas) in the vicinity of Earth's surface. Using this value for the geomagnetic field, representative values for protons and electrons are *given in Table 1.1* [Editors note: 1 wb/m^2 = (tesla = 10^4 gauss = 10^9 gamma].

1.7 Particle Drifts

The equations from the previous section hold in a uniform magnetic field, and are good approximations for anything approaching a uniform magnetic field. However, it should be stressed that if there were an external force, or if the magnetic field were not uniform, the charged particles would not follow perfectly circular paths. While the radius and period of gyration would both be given to a reasonable degree of accuracy by equations 1.26 and 1.28 the particles would drift through the magnetic field in addition to performing the basic circular motion that has been discussed. To clarify this point, four cases will be considered: (1) an external force which is independent of charge, (2) an external force which is dependent on charge, (3) a non-uniform magnetic field, and (4) a curvature in the magnetic field geometry.

For case (1) let us assume that a force (such as gravity) lies in the plane of the page, directed downward. This means that during part of the particle's path both the external force and the magnetic force are directed toward the bottom of the page. This increases the radius of gyration to a value greater

Table 1.1 Some representative values for single particle motion in the geomagnetic field for which $B = 10^{-5}$T.

	Proton	Electron
q (coul)	1.6×10^{-19}	-1.6×10^{-19}
m (kg)	1.67×10^{-27}	9.1×10^{-31}
E (MeV)	1	0.5
r_c (m)	1.5×10^4	7.5×10^1
ω_c (sec^{-1})	958	1.76×10^6

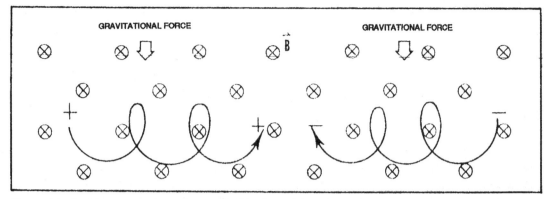

Figure 1.1 In this example the uniform B field is directed into the page and the external force (gravity) acts downward, resulting in positive particles drifting to the right and negative particles to the left. The difference between the positive and negative particle motions is a net positive current to the right *(after Carpenter et al., 1978)*.

than that for magnetic force alone. For the other part of the particle's path, the external force and the magnetic force are in opposition. This decreases the radius of gyration; the net result is to cause the particle to drift as shown in Figure 1.1. In general, the particle's drift velocity can be written as

$$\vec{v}_d = \frac{\vec{F} \times \vec{B}}{qB^2} \qquad (1.31)$$

where $\vec{F}$ is the external force acting on the particle with charge q which is within a magnetic field $\vec{B}$. In the case of gravity, $\vec{F} = m\vec{g}$. Notice that the particles drift direction depends on the *sign* of the charge which must be included with q. The result is that electrons and positive ions gravitationally drift in opposite directions, and the resulting charge separation produces a net positive drift current to the right which, in turn, feeds back and modifies $\vec{B}$.

For case (2), let us assume that an electric field, $\vec{E}$, lies in the plane of the page, directed toward the bottom of the page. This produces a force whose direction is charge dependent. Applying an analysis similar to that for case (1), we find again that the particles drift (see Figure 1.2). The drift velocity becomes

$$\vec{v}_E = \frac{q\vec{E} \times \vec{B}}{qB^2} = \frac{\vec{E} \times \vec{B}}{B^2} \qquad (1.32)$$

where the external force of equation 1.31 is $\vec{F} = q\vec{E}$ and the resulting $\vec{E} \times \vec{B}$ drift is *charge* independent. Therefore, electrons and protons drift in the same direction, and at the same speed, resulting in no net current.

For case (3), let us assume a non-uniform magnetic field, with the greatest field strength nearest the bottom of the page. Again applying an analysis similar to the above, we find the particles drift as shown in Figure 1.3. In this case the external force can be written as

$$\vec{F} = -\tfrac{1}{2}mv_\perp^2 \frac{\vec{\nabla}BN}{B}$$

where $v_\perp$ is the component of the particle's velocity around (perpendicular to) B. Therefore, the so-called gradient drift becomes

$$\vec{v}_{grad} = \tfrac{1}{2}\frac{mv_\perp^2}{q}\frac{\vec{B} \times \vec{\nabla}\vec{B}}{B^3}. \qquad (1.33)$$

It is customary to write equation 1.33 in terms of the particle's magnetic dipole moment μ. In its simplest terms, the magnetic moment of a current loop is just the current in the loop times the area of the loop. The gyrating particle, going around once per time $2\pi/\omega_c$ carries an equivalent current $q\omega_c/2\pi$. Therefore, the magnetic moment is given by

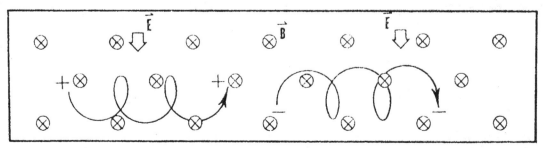

Figure 1.2 Cross B field drift due to an external electric field. In this example, the force is charge dependent and the resulting drift is the *same* for both positive and negative particles. Therefore, there is *no* net current produced. Once again, B is directed into the page *(after Carpenter et al., 1978)*.

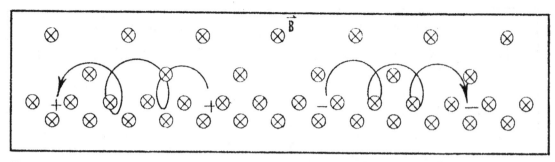

Figure 1.3 Charged particle drifts due to a non-uniform magnetic field. The magnetic field is again directed into the page and the field strength increases toward the bottom of the figure. The resulting drift motion (positive to the left and negative to the right) produces a net positive current to the left *(after Carpenter et al., 1978)*.

$$\mu = \frac{q\omega_c}{2\pi} \cdot \pi r_c^2 = \frac{1}{2}\frac{mv_\perp^2}{B} \qquad (1.34)$$

where we have used both equations 1.26 and 1.28. As we will see in the next section, μ is nearly always a constant for space science applications. The gradient drift depends on the charged particle's perpendicular kinetic energy and not just on the velocity. Thus, electrons and protons that have the same *kinetic temperature* actually undergo gradient drifts at the same rate, but in opposite directions, even though the electrons have *much greater* thermal velocities.

The physical picture of the gradient drift is that the gyroradius is smallest at the bottom of the gyro-orbit because B is increasing in that direction. For the magnetic field configuration of Figure 1.3, a positive particle is moving along the positive x-axis at the bottom of the orbit (smallest radius of curvature), and along the negative x-axis at the top of the orbit where the radius of curvature is the largest. The net result is that the positive particle moves to the left. Conversely, electrons will drift to the right, and consequently, gradient drifts tend to produce net electrical currents.

Figure 1.3 is similar to Earth's magnetic field in that the further one goes from the Earth, the weaker the magnetic field becomes. Also, the figure's orientation is such that it provides essentially the view one has when looking northward along the lines of force of Earth's magnetic field. The resulting gradient drift current has important consequences for the near Earth space environment (see Chapter 5).

For case (4), let's suppose that the magnetic field is uniform except for some small variation which gives the total field some small curvature in a particular direction. There are no external electric fields. As the gyrating particles move along a curved field line, some external force must act on the particle to make it turn and follow the field line geometry. If there are no external electric fields, this turning force must be provided by the magnetic field. As the charged particles follow $\vec{B}$, then a force of

$$\vec{F} = \frac{mv_\parallel^2}{R}\hat{R} \qquad (1.35)$$

is needed to turn the particle where $v_\parallel$ is the particle velocity parallel to B and R is the radius of *curvature* (not gyroradius) of the magnetic field lines. Therefore, using equation 1.35 in equation 1.31 we find.

$$\vec{v_c} = \frac{mv_\parallel^2 \vec{R} \times \vec{B}}{qR^2B^2} \qquad (1.36)$$

where the explicit appearance of q tells us positive and negative particles will undergo curvature drift in opposite directions, and thus produce a net electrical current. Also notice that the curvature drift depends on the particle's parallel component of kinetic energy. Thus, electrons and protons with the same parallel energy drift at the same speed, but in opposite directions, even though the electrons have a much larger thermal velocity due to their smaller mass.

The theory of single particle drifts (sometimes called the "guiding-center" theory) produces currents which are much different from the Ohmic currents (section 1.2) that we are accustomed to in the laboratory and even in our everyday electrical appliances. Ohmic currents are collisionally dominated, and it is in these collisions that the charge carriers transfer and lose energy. The drift currents presented in this section are non-collisional and are due to the reaction of charged particles to variations in the electromagnetic field structure. Normally, we treat these drift currents as non-lossy but, in fact, there are some minor radiation losses due to the particle accelerations associated with changes in gyro orbit sizes and directions. The particle drift picture is particularly powerful in low density plasmas, but becomes progressively less useful as the probability of particle collisions becomes more likely. At times, both ohmic and drift currents are important in some geophysical applications. For example, in the ionosphere ohmic currents dominate, but even these ohmic currents are affected by changes in the magnetic field geometry due to drift currents in the magnetosphere.

1.8 Magnetic Mirroring

It is extremely important to any individual who is interested in space travel, or the space environment, to have some understanding of the way a planetary magnetic field traps charged particles. This will require the development of more information on the motion of charged particles in magnetic fields. Some of this information has been already introduced in earlier sections. In particular, cyclotron and drift motion were both discussed in adequate detail. This leaves the mechanism of mirroring, and the invariants of motion, to be covered. In the preceding sections we've already discussed that when a charged particle moves with some velocity $\vec{v}$ in a magnetic field $\vec{B}$, a force is exerted on the charged particle. The direction of the force is perpendicular to $\vec{B}$, and the magnitude of this force is proportional to both the magnitude of $\vec{B}$ and the component of the velocity which is at right angles to $\vec{B}$. This suggests that charged particle motion in the direction of $\vec{B}$ is unaffected by the magnetic fields, however, this conclusion is wrong in the cases of converging and diverging magnetic fields.

In Figure 1.4, for example, a cross section is shown of a proton's path in a diverging magnetic field. The proton is shown performing the basic cyclotron motion such that it is entering the page near the top of the figure, and leaving the page near the bottom. As is seen, the force acting at right angles to the $\vec{B}$ vector does not lie in the plane of circular motion of the charged particle. Instead, the force vectors form a cone whose apex is in the direction of decreasing magnetic field strength. Thus, there is a net displacement force in the direction of weaker field strength. This same result holds true for an electron in a converging field.

This means that a charged particle spiraling in a magnetic field will experience a force along the field lines, in the direction of weaker field strength. When the magnitude and duration of the force are sufficient to actually cause the

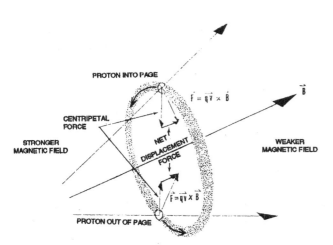

Figure 1.4 Forces acting on a gyrating proton in a diverging magnetic field. Notice that the net displacement force is directed away from the region of stronger (converging) field strength, and toward the region of weaker (diverging) field. Therefore, as a charged particle moves into a region of converging B, a force acts to slow the particle down, and if the force is strong enough and acts long enough, the charged particle can change direction (mirror point) along the magnetic field lines.

charged particle to reverse direction of motion along the line of magnetic force, the effect is known as *mirroring* and the location of the particle's path reversal is known as the mirror point for that particle.

Surprisingly, the kinetic energy of a magnetically trapped particle is essentially a constant throughout its path. That is, the Lorentz force ($\vec{F} = q\,\vec{v} \times \vec{B}$) is the only external force acting on the mirroring particle, and the Lorentz force always acts perpendicular to the direction of motion which results in no net work being done. Thus, the kinetic energy of the particle does not change, only the direction of motion changes. Once again, we are assuming that radiation losses (Bremsstrahlung due to the particle accelerations associated with changes in the gyro-orbit size and direction) are small.

So we now know that charged particles can be trapped in our geomagnetic field. Their basic motion is circular, with a superimposed longitudinal drift around the Earth, and a latitudinal reflection (or "bounce") between mirror points at high latitudes. The bounce period is on the order of seconds for both protons and electrons (see Figure 5.5).

Further information on the behavior of trapped particles can be obtained by consideration of three assumed invariants of motion: the magnetic moment invariant, the action integral invariant, and the flux integral invariant.

1.9 First Adiabatic Invariant/Magnetic Moment Invariance

The amount of magnetic flux, Φ, enclosed by the cyclotron path of the charged particle is

$$\Phi = \int \vec{B} \cdot d\vec{A} \qquad (1.37)$$

where $d\vec{A}$ is a unit area intersected by $\vec{B}$. If we assume that B does not vary throughout the cross sectional area dA, then equation 1.37 becomes

$$\Phi = \pi r_c^2 B \qquad (1.38)$$

where r_c is the cyclotron radius and B is the magnetic field magnitude. As we saw in section 1.6 the centripetal force required to turn the particle in the gyro-orbit is simply the Lorentz force, and therefore the gyroradius is given by

$$r_c = \frac{m}{q}\frac{v_\perp}{B} \qquad (1.39)$$

where $v_\perp$ is again just the velocity component perpendicular to B. Using equations 1.39 and 1.34, the flux becomes

$$\Phi = \frac{2\pi m}{q^2}\mu. \qquad (1.40)$$

More insight into the behavior of Φ can be gained by relating the particle's kinetic energy to the pitch angle which is the angle between the magnetic line of force and the direction of motion of the particle, as shown in Figure 1.5. In terms of the pitch angle α, $v_\perp$ is seen to be

Figure 1.5 Pitch angle α, the angle between the magnetic line of force and the direction of the charged particle's motion.

$$v_\perp = v \sin \alpha \tag{1.41}$$

and the magnetic moment in equation 1.34 becomes

$$\mu = \tfrac{1}{2} \frac{mv^2}{B} \sin^2 \alpha. \tag{1.42}$$

Since we've already shown that the work done on the particle is zero (hence $\tfrac{1}{2}mv^2$ = constant), then the magnetic moment depends on how the pitch angle changes with B. In order to determine this functional relationship we need to turn our attention to Φ.

Faraday's law of induction (see equation 1.6) says that the induced electromotive force (Emf) is equal to the rate at which the flux through the circuit is changing or

$$\text{Emf} = -\frac{d\Phi}{dt} \tag{1.43}$$

where the minus sign indicates the direction of the induced Emf. Physically, Emf is the ability of the fields to do work which, in turn, changes the energy. Since the energy of the particle is constant, the above argument says that the change of the particle produced flux must be zero. That is,

$$\frac{d\Phi}{dt} = 0 \tag{1.44}$$

which means the magnetic flux (Φ) enclosed by the cyclotron path of the charged particle is a constant. Therefore, from equations 1.40 and 1.42 we find

$$\mu = \tfrac{1}{2} mv^2 \frac{\sin^2 \alpha}{B} = \text{constant.} \tag{1.45}$$

So the pitch angle α must increase with increasing field strength. The change of pitch angle, in turn, is a change in the distribution of the particle's kinetic energy. Thus, the particle's kinetic energy increases in the cyclotron direction (at right angles to $\vec{B}$) and decreases in the direction parallel to the magnetic field. The mirror point is reached when all of the particle's kinetic energy is along the cyclotron direction.

Equation 1.45 is usually referred to as the *first adiabatic* invariant; μ remains nearly constant during one gyro-orbit, even though B may be changing slowly. The invariance of μ is violated when the frequency characterizing the rate of change of B (as seen by the particle) is faster than the cyclo-

tron frequency. Under such conditions, equation 1.38 is invalid.

Another relation of interest can also be obtained here by noting that at the particle's mirror point (denoted by subscript m), the pitch angle goes to $\pi/2$ radians, and equation 1.40 becomes

$$\Phi = \left(\frac{2\pi m}{q^2}\right) \left(\frac{mv^2}{2}\right) \frac{1}{B_m} \tag{1.46}$$

Dividing equation 1.46 by equation 1.40 and eliminating μ by substituting in equation 1.42 yields

$$\sin^2 \alpha = \frac{B}{B_m}. \tag{1.47}$$

This relates the pitch angle α to the magnetic field strength B at any point in space.

1.10 *Second Adiabatic Invariant/ Longitudinal Invariant*

The second adiabatic invariant says that the integral of the parallel momentum over one complete bounce between mirror points is constant. That is,

$$J = \int_{s_1}^{s_2} 2mv_\parallel \, ds = \text{constant} \tag{1.48}$$

or using the conservation or energy and equations 1.41 and 1.47 gives us

$$J = \int_{s_1}^{s_2} 2mv \left(1 - \frac{B}{B_m}\right)^{1/2} ds = \text{constant.} \tag{1.49}$$

Here s_1 and s_2 are the mirror points (limit of particle's path) and the s coordinate, directed along the field line, is zero at the point of minimum B. B at the mirror point is denoted by B_m. The derivation of equation 1.48 is beyond the scope of this text and involves finding the action integral using Lagrangian mechanics. The interested reader is referred to Chen (1974). This invariant remains valid as long as the field does not change appreciably during the time of one bounce.

An application of this invariant is found in the slow drift of geomagnetically trapped particles. Transit times from one mirror point to the other are on the order of a second. In addition, our earlier discussion of particle drifts implies that the trapped particles will *drift* azimuthally since the geomagnetic field lines are curved and there is a magnetic field gradient normal to the direction of B. Electrons drift from west to east and protons from east to west, and the particles trace out a surface of rotation known as the longitudinal invariant surface. (A 40 keV electron takes about an hour to drift completely around the Earth.) If the geomagnetic field were completely symmetric, a particle would eventually drift back to the same magnetic field from which it originated. However, we know that the geomagnetic field is asymmetric (see Chapter 5) and it is reasonable to ask if the particles will ever drift back to the same line of force (magnetic field line)? The first adiabatic invariant indicates that B at the mirror point remains the same for particles with a given energy. The conservation of J determines the length of the field line between mirror points.

Another application of J is when the mirror points are no longer stationary but move toward one another very slowly so as not to invalidate the invariance of J. Another way to write equation 1.48 is

$$J = \oint mv_\parallel ds = \text{constant} \qquad (1.50)$$

where the circle on the integral symbol implies that the integral is over the closed path which starts and ends at S_1. If the mirror points move closer together (the path length decreases), then in order for J to remain constant, $v_\parallel$ must increase. Physically, the easiest way to think of the process is to imagine throwing a ball at a wall which is moving toward you. The ball bounces off the wall with more speed than it had approaching the wall since the ball gains energy from the moving wall. Similarly, as the mirror points more steadily (but slowly) toward one another, the trapped particles gain parallel velocity and the energy source is the external mechanism which is modifying the mirror points. This mechanism has been suggested as a possible source for the very energetic cosmic ray particles. Since magnetic fields exist in space, it is plausible to assume that there are regions in which charged particles may become trapped. Moreover, the trapping regions may not be static and the trapped particles could be accelerated if the ends of the trapping region approach one another. Conversely, particles will lose energy if these mirror regions were separating. This cosmic ray model was originally proposed by Fermi, but the model has the limitations that as $v_\parallel$ increases, the pitch angle decreases so that, ultimately, the particle will eventually escape from the magnetic bottle. Thus, for the Fermi mechanism to be applicable, some additional feature (such as particle collisions) is needed to divert some of the increasing particle energy into the perpendicular direction in order to keep the pitch angle large enough for confinement.

1.11 Third Adiabatic Invariant/Flux Invariance

The third, and weakest (the least likely to be continually valid), invariant of particle motion is that the magnetic flux Φ enclosed by the charged particle longitudinal drift ring must be a constant. This is really a statement of Faraday's law of induction applied to a constant energy system. As $\vec{B}$ varies (with longitude), the particle will stay on a surface such that the total number of field lines enclosed remains constant. This invariant has few applications because most fluctuations of $\vec{B}$ occur on time scales which are short compared to the drift period.

1.12 Limitations on the Invariants

It is important at this time to review the space and time requirements on the three invariants discussed. The only invariant with a space requirement is for μ. Simply, the magnetic moment μ is constant when there is little change in the field's strength over the cyclotron path. Therefore

$$\left| \frac{\nabla B}{B} \right| \ll \frac{1}{r_c}. \qquad (1.51)$$

However, all three invariants had the time requirement that only small changes in B can be permitted during the orbit period; otherwise, the particle will not retain its path or energy. Symbolically,

$$\left| \frac{1}{B} \frac{\partial B}{\partial t} \right| \ll \frac{1}{\tau} \qquad (1.52)$$

where τ is the orbit period. For example, for the magnetic moment μ invariance, τ is approximately 10^{-3} sec for protons, and 10^{-6} sec for electrons; for J invariance, τ is approximately 1 sec; for the flux invariance Φ, τ is on the order of minutes. In practice, if the right-hand side of equation 1.52 is a factor of 100 or greater than the left-hand side, then the invariant (with respect to the time constant) is expected to hold.

In a time varying magnetic field, the Φ invariant would fail first, the J invariant second, and the μ invariant last. The rate of time variation would determine when a particular invariant would fail. In this regard, it should be seen that geomagnetic storm fluctuations occur on the order of minutes, throwing grave doubts on the invariance of Φ (at least during geomagnetic storms).

1.13 Magnetic Energy and Pressure

In elementary physics courses one is exposed to the concept that a magnetic field is really stored energy. The simplest example is an inductor in common oscillating electrical circuits. In such a circuit, the amount of energy stored in B depends on the amount of current in the circuit plus the physical dimensions of the inductor itself.

Upon closer examination of the physical interactions of B, we find that it is very useful to develop the notion of magnetic momentum density and magnetic momentum flux. Attributing momentum density and flux to the electromagnetic field is not unreasonable, since the field is made up of photons, and no one objects to attributing momentum to particles. Also, it is necessary to attribute a momentum density to the electromagnetic field in order to satisfy the ideal of momentum conservation in a closed system.

A complete analysis shows that the magnetic momentum flux is a tensor and consists of two parts: magnetic tension and magnetic pressure (or energy density). The magnitude of the magnetic tension is B^2/μ_o, acting in the direction of the magnetic field. The second term is an isotropic pressure of strength $B^2/2\mu_o$. Another way of thinking of the magnetic pressure is as energy density stored at any point where the magnetic field strength is B.

It is very useful to compare the magnetic pressure to the plasma gas pressure because in many cases

$$P + \frac{B^2}{2\mu_o} = \text{constant.} \qquad (1.53)$$

This equation is simply saying that the total pressure in a closed system is a constant and it is generally true as long as field gradients (terms having the form $\vec{B} \cdot \nabla \vec{B}$) are small and the plasma is stationary. In cases where the fluid is flowing, we may need to include the dynamic pressure effects and equation 1.53 becomes

$$\rho v^2 \cos^2\psi + P + \frac{B^2}{2\mu_o} = \text{constant} \qquad (1.54)$$

where ψ is the angle between the velocity vector and the normal to the surface on which the plasma is acting. The angle ψ is particularly important at shock boundaries (see Chapter 5) where the boundary shape is determined by balancing the total pressure inside the shock with the total external pressure.

It is convenient to define β as the ratio of the plasma and dynamic pressure to the magnetic pressure

$$\beta = \frac{P + \rho v^2}{B^2/2\mu_o} \qquad (1.55)$$

where the plasma gas pressure, P, is assumed to be isotropic. A "high beta" plasma ($\beta \gg 1$) is one which is controlled principally by the plasma gas dynamics and a "low beta" plasma ($\beta \ll 1$) is dominated by the intrinsic magnetic field.

1.14 Diamagnetism

A plasma will resist both the intrusion of any applied external magnetic fields and any change to an embedded (within the plasma) magnetic field. To demonstrate this "diamagnetic" character of a plasma, we need to return to section 1.6 and review the gyromotion of protons and electrons about a magnetic field line. If the magnetic field is directed out of the page, then the proton will rotate clockwise and electrons will rotate counterclockwise. Therefore, the net effect of many gyrating particles is to produce a circular current which, in turn, induces a magnetic field which opposes the applied field. The plasma is trying to eliminate the applied magnetic field.

The magnitude of the induced field can be estimated from Ampere's law:

$$\nabla \times \vec{B}_{ind} = \mu_o \vec{J}_{ind} = \mu_o Ne(\vec{v_i} - \vec{v_e}) \qquad (1.56)$$

or in terms of scale lengths

$$\frac{B_{ind}}{r_c} \sim \mu_o Nev_\perp \qquad (1.57)$$

where N is the particle density, r_c is the gyroradius, e is the charge on an electron, and $v_\perp$ is the gyrovelocity about B. Therefore, the magnitude of the induced field to the applied field B is

$$\frac{B_{ind}}{B} \sim 2\mu_o \frac{NK_\perp}{B^2} \qquad (1.58)$$

where we have used both equations 1.26 and 1.57, and $K_\perp = \frac{1}{2}mv_\perp^2$. The intensity of the induced field depends on the rate of change of the applied field (Faraday's law) because the induced electric field can drive the particles "harder" and thereby increase their kinetic energy.

Induced magnetic fields are particularly important in the interaction of the solar wind with the ionospheres of unmagnetized planets, such as Venus. Because of this interaction, the induced planetary magnetic field of Venus produces a bow shock to deflect a large portion of the solar wind around the planet. Similar induction type magnetospheres might exist on some moons (with atmospheres) of Jupiter and Saturn. Titan is a good example. In this case, the induced field is not due to the interaction of the solar wind, rather it results from the interaction of Titan's ionosphere with the main magnetic field of Saturn.

1.15 Electromagnetic Radiation

Like particle radiation, electromagnetic radiation can produce ionization, excitation, and dissociation in matter. This is true even though the electromagnetic packets of energy (called photons) lack electric charge—the lack of electric charge only makes them tend to have a lower probability of interaction with matter. This lower interaction probability has its own importance however. It means that the high energy photons cause less specific ionization per unit volume than do charged particles. Therefore, radiation damage is produced through a larger volume of material.

There are a number of ways that a photon can interact with matter, including Rayleigh scattering, pair production, Compton scattering, and photoelectric absorption. A short summary description of each of these phenomena is given below. The rate at which photons are removed from an incident beam depends on the type and density of matter through which the beam is passing. Mathematically this is usually represented as

$$I = I_o \exp\left[-\left(\frac{u}{\rho}\right)(\rho x)\right] \qquad (1.59)$$

where I is photon intensity at distance x into the matter, I_o is the intensity of photons incident on surface of the matter, ρ is the mass density of matter, and u is the macroscopic probability of photon removal. The term (u/ρ) is referred to as the *mass absorption coefficient*, and it is primarily the sum of the mass absorption coefficients for pair production, Compton scattering, and the photoelectric effect.

Rayleigh scattering is the elastic scattering of a photon off of a "bound" electron (i.e., an electron that is part of a neutral atom) with insignificant loss of photonic energy. It is sometimes referred to as coherent scattering, and its wavelength dependency causes short wavelength light to be scattered more than long wavelength light. This results in Earth's atmosphere appearing blue, and the Sun near the horizon appearing red at sunset.

Pair production is the conversion of a photon into an electron and a positron. Conservation of momentum requires this process to occur in the vicinity of an atomic nucleus, the heavier the nucleus the more probable the interaction, provided the photon possesses enough energy to form the electron and positron masses, plus the recoil energy of the momentum-conserving nucleus. Since this latter energy is usually very small, we generally neglect it. Any photon energy beyond the minimum necessary for the interaction to occur will appear as kinetic energy of the electron and positron. The energy equivalent of the rest mass of an electron is 0.511 MeV. The positron energy equivalent is the same, so a photon must possess at least 1.02 MeV for pair formation to occur. This energy is provided by a photon with a wavelength of 1.23×10^{-2} Å (1.23×10^{-12} m), which is effectively outside the range of the solar emission spectrum—it is in the gamma particle range of the electromagnetic spectrum.

Compton scattering is the elastic scattering of photons by free electrons—that is, by electrons that behave as if they were free from atomic bonds. In this interaction, the photon retains some energy, the amount of which depends upon the initial photon energy and on the angle of scatter.

Photoelectric absorption results in the complete annihilation of the photon, as did pair production. In the photoelectric case, however, the initial photon energy is used to free an electron from an atom. Any leftover energy appears as kinetic energy of the electron.

1.16 Ionization and Excitation

Two processes are responsible for most of the observed natural atmospheric ionization. The first process is photoionization where an atom (or molecule) absorbs a photon of sufficient energy to free a bound electron. The second process is collisional ionization where during a collision, a neutral atom absorbs enough energy to free a bound electron. In either case, the result of the ionization process is a positive ion and at least one free electron. Another type of ionization which occurs naturally in the upper atmosphere involves

negative ions. In the case, a neutral atom or molecule with a large electron affinity "attracts" a free electron which attaches itself to the neutral, thereby giving the atom an excess negative charge. This excess electron is only loosely attached to the atom and the bond is easily broken. Such negative ions are critically important in the ionospheric chemistry of the ionospheric D region (see Chapter 7 and following).

A free electron will eventually neutralize a positive ion and the captured electron is usually in an excited (above the ground state) energy level of the newly formed neutral atom. Typically, the electron quickly drops down to the lowest available energy state in one or more steps, and with each step transition a photon is radiated whose energy is simply equal to the difference between the energy of the upper and lower energy states of the transition. The transition levels for a hydrogen atom are shown in Figure 1.6. Notice that specific sets of transitions are uniquely identified (e.g., Lyman series, Blamer series, etc.). However, another sequence of events can occur where a captured electron can be de-excited to a lower energy state without the emission of a photon. These so-called "radiationless transitions" occur if the neutral, with the electron still in the excited state, suffers a collision which triggers the downward transition of the excited electron. However, instead of producing a photon, the excess energy released by the transition is transferred to and carried off as additional kinetic energy by the other objects involved in the collisions. These other objects may be another neutral or even a charged particle. In some cases where an

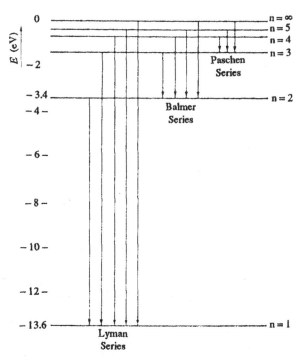

Figure 1.6 Energy level diagram for hydrogen.

excited neutral collides with a positive ion, an electron can be transferred to the positive ion as well as excess kinetic energy. Notice that in such "charge exchange" collisions, the total charge is conserved. The total number of ions doesn't change, even though the actual identity of individual ions may change. In Chapter 7 we will see that charge exchange reactions are crucial in the formation of the F region of the ionosphere.

Some representative ionization energies and the associated wavelengths for photoionization are:

$$NO : E = 9.25 \text{ eV } (\lambda = 1340 \text{ Å})$$
$$O_2 : E = 10.08 \text{ eV } (\lambda = 1027 \text{ Å})$$
$$H_2O : E = 12.60 \text{ eV } (\lambda = 985 \text{ Å})$$
$$H : E = 13.60 \text{ eV } (\lambda = 912 \text{ Å})$$
$$O : E = 13.61 \text{ eV } (\lambda = 911 \text{ Å})$$
$$N : E = 14.54 \text{ eV } (\lambda = 853 \text{ Å})$$
$$H_2 : E = 15.41 \text{ eV } (\lambda = 804 \text{ Å})$$
$$N_2 : E = 15.58 \text{ eV } (\lambda = 796 \text{ Å})$$
$$He : E = 24.58 \text{ eV } (\lambda = 504 \text{ Å})$$

where the energies E are expressed in electron volts and the wavelengths are in angstroms (1 Å = 10^{-10}m).

Thus far, our discussion has been confined to photoionizations. However, within the auroral zone precipitating electrons and protons from the magnetosphere are the principal source of ionizations. These collision induced ionizations can be symbolized by

$$M + e \rightarrow M^{+*} + e + e$$

or

$$M + H^+ \rightarrow M^{+*} + H^+ + e$$

or

$$M + H^+ \rightarrow M^{+*} + H^*$$

where M is the atmospheric neutral atom or molecule, M^{+*} is the resulting ionized atmospheric element in an excited electronic state, e and H^+ on the left side of the equations represent the precipitating electrons and ions, and H^* represents the case where the electron freed by the collision recombines with the proton to form a hydrogen atom in an excited electronic state. In the case of the auroral zone, the most likely target atoms (M) are molecular (and atomic) oxygen and nitrogen. The actual rate of ionization depends on both the collision cross sections and number densities of the atmospheric elements.

Experiments show that precipitating protons, or electrons, produce about one ionization per 36 eV of their initial energy. Referring to the ionization potential chart above, we see that for most atmospheric elements the ionization energy is about 15 eV, and therefore, about 60% of the 36 eV used in the ionization reaction goes into excitation or translational energy of the target element.

Some of the ionized atmospheric elements return to their ground states by collisions while others relax to their ground states through photon emission. Therefore, by measuring the emission rates, we should be able to determine the extent of collision ionizations. For example, theory and experiment show that the emission from one of the N_2^+ molecular bands is independent of the energy of the precipitating electrons, and that for every 25 ion-electron pairs formed by collisions at auroral altitudes, one ion relaxes by emitting a photon at 3,914 Å; the others lose their energy via collisions.

Emissions from proton induced ionizations is more complicated because the photoemission/ion-electron pair ratio is dependent on the incident proton energy. For example, at an energy of 10 KeV the 3,914 Å ratio is 38, but at an energy of 3 KeV the ratio increases to about 120. Similarly, different photoemission/ion-electron ratios exist for each molecular band emission line, and therefore, an observer on Earth (or sensors on board a spacecraft) can estimate the auroral ionization rates by monitoring specific emission lines. Unfortunately, these measurements are complicated by secondary electron ionizations.

As mentioned above, about 60% of the 36 eV used in the collision ionization can be carried away by the electrons created during the ionization process. These secondary electrons can then become sources of both additional ionization and excitation. Theory and experiment must then be used to correct the photoemission/ion-electron ratios for secondary electron ionizations.

1.17 Summary

In this chapter we have briefly reviewed some of the underlying principles of plasma physics. We have seen that in many cases a plasma acts as a fluid and the bulk flow equations are adequate description of the macroscopic properties of the plasma. At other times, we are concerned with the microscopic properties of the plasma, and at these times, we need to look at the single particle motion of the individual charges making up the plasma. We also discovered that certain properties of the single particle motion are invariant when the background magnetic field changes very slowly. These adiabatic invariants simplified our description of magnetic bottles and particle trajectories in the Earth's magnetosphere. Another important plasma parameter was the term β which is the ratio of plasma energy density to the magnetic energy density. Diamagnetism was shown to be a basic and critically important characteristic of a plasma. Finally, we found that electromagnetic radiation is very important in the production and maintenance of the ionosphere, a natural atmospheric plasma.

1.18 References

Alfven, Hannes. 1981. *Cosmic Plasma*. Astrophysics and Space Science Library, vol. 82, D. Reidel Publishing Co., Boston.

Bostrom, R. 1973. Electrodynamics of the Ionosphere, in *Cosmical Geophysics*, A. Egeland, O. Holter, and A. Omholt, eds. Universitetsforlaget, Oslo.

Boyd, T. J. M., and Sanderson, J. J. 1969. *Plasma Dynamics*. Barnes and Noble, Inc., New York.

Carpenter, D. G. et al. 1978. *Introductory Space Physics*. U.S. Air Force Academy, Col.

Chen, F. F. 1974. *Introduction to Plasma Physics*. Phenum Press, New York.

Cloutier, P. A., Tascione, T. F., Daniell, R. E., Taylor, H. A., and Wolff, R. S. 1983. Physics of the Interaction Flow/Field Models, in *Venus*, (Hunter, Colin, Donahue, and Moroz, eds.) University of Arizona Press, Tucson.

Hargreaves, J. K. 1979. *The Upper Atmosphere and Solar-Terrestrial Relations*. Van Nostrand Reinhold Co., New York.

Omholt, A. 1973. Particle Precipitation: Ionization and Excitation, in *Cosmical Geophysics*. A. Egeland, O. Holter, and A. Omholt, eds. Universitetsforlaget, Oslo.

Piddington, J. H. 1981. *Cosmic Electrodynamics*. Krieger Publishing Co., Malabar, Fla.

Reitz, J. R., and Milford, F. J. 1967. *Foundations of Electromagnetic Theory*. Addison-Wesley Publishing Co., Reading, Mass.

Wolf, R. 1979. *Plasma Physics*, unpublished book/class notes. Rice University, Houston, Tex.

1.19 Problems

1-1.   If the equation of motion for a charged particle is

$$m\frac{d\vec{v}}{dt} = q\,\vec{v} \times \vec{B}$$

a.   Show that if $\vec{v}$ is perpendicular to $\vec{B}$, the particle trajectory is a circle, and that the gyroradius is given by

$$r_c = \frac{mv}{Be}$$

(do not start with the equation of centripetal acceleration because this equation assumes circular motion—you must *show* that the trajectory is circular).

b.   Compute the gyroradius for a solar wind proton and electron. The solar wind speed is about 400 km/sec and B is about 5 gammas. (1 gamma = $1\tau = 10^{-9}$ tesla.)

1-2.   Starting with Ampere's law

$$\nabla \times \vec{B} = \mu_o\,\vec{J} + \mu_o\epsilon_o\frac{\partial \vec{E}}{\partial t}$$

show that when $\vec{J} = 0$, the electric displacement current $(\mu_o\epsilon_o\,\partial \vec{E}/\partial t)$ implies that $\vec{B}$ (and/or $\vec{E}$) moves through space as transverse waves.

1-3.   In many space related situations the electric displacement current is ignorable (sometimes called steady state conditions). Over what period of time can we consider the solar wind to be in steady state? How about the ionosphere? Take the conductivity of the solar wind to be nearly infinite, and the conductivity of the ionosphere to be 10^{-3} mho/m.

1-4.   Under steady state conditions the magnetic fields can still move through space by the process of diffusion. Derive an expression for the diffusion time. Explain your method.

1-5.   a.   An important property of a highly conducting plasma is that the magnetic flux becomes "frozen-in" the plasma flow. Starting with equation 1.6 (and using Stokes' theorem) show that

$$\frac{d\Phi_B}{dt} = -\int \nabla \times (\vec{E} + \vec{v} \times \vec{B}) \cdot d\vec{A}$$

and for "frozen-in-flux" conditions

$$\nabla \times (\vec{E} + \vec{v} \times \vec{B}) = 0.$$

b.   Show that for interplanetary conditions (high conductivity) frozen-in flux implies

$$(\vec{E} + \vec{v} \times \vec{B}) = 0.$$

c.   When frozen-in-flux conditions are valid, describe the motion of a charged particle moving with respect to the lab frame at a velocity $\vec{v} = (\vec{E} \times \vec{B})/B^2$.

1-6.   Describe the motion of an electron and a proton under the influence of an external force $(\vec{F}_{ext})$ which acts transverse to a uniform magnetic field.

1-7.   In a magnetic mirror machine, compute the magnitude and direction of the forces as a charged particle approaches the mirror point.

1-8.   Show that in a mirror machine (no collisions) particles with

$$\mu < \frac{mv^2}{2\,B_{max}}$$

can escape down the field lines (μ = magnetic moment).

1-9.   Show that the first adiabatic invariant implies that the particle's angular momentum is also conserved.

1-10.   a.   Assume that in a partially ionized gas the collisions between neutrals and charges dominate. In the limit of small plasma accelerations, show that (see section 1.2)

$$\sigma_P = \frac{N_e e^2}{M_e}\left(\frac{\nu_{en}}{\nu_{en}{}^2 + \omega_{ce}{}^2} + \frac{M_e}{M_i}\frac{\nu_{in}}{\nu_{in}{}^2 + \omega_{ci}{}^2}\right)$$

$$\sigma_H = \frac{N_e e^3 B}{M_e{}^2}\left(\frac{-1}{\nu_{en}{}^2 + \omega_{ce}{}^2} + \frac{M_e{}^2}{M_i{}^2}\frac{1}{\nu_{in}{}^2 + \omega_{ci}{}^2}\right)$$

$$\sigma_\parallel = \frac{N_e e^2}{M_e \nu_{en}} + \frac{N_e e^2}{M_i \nu_{in}}$$

b. Using the results of part a, show why an equatorial electrojet forms. Does this mechanism work near the poles? Explain. (Hint: At the equator, assume there is no vertical current.)

1-11. Starting with the continuity equation

$$\frac{\partial N_e}{\partial t} + \nabla \cdot (N_e \vec{v}) = 0$$

and the equation of motion (with an electric field as the only force), show that plasma frequency is given by

$$\omega_p = \sqrt{\frac{e^2 N_e}{M_e \epsilon_o}}.$$

Chapter 2

Solar Physics

2.0 Introduction

The Sun is a star of only average luminosity and surface brightness, and one of about 10^{11} in our galaxy. Our interest lies, of course, in its proximity and consequent important effects on our environment. This relative closeness also allows the Sun to be studied in vastly greater detail than the next nearest star and so its optical, thermal radio, x-ray, and particle radiations are monitored continuously.

The Sun has an average radius of 696,000 km (about 109 times the radius of the Earth), and it would take over 1 million Earths to fill the volume of the Sun. The Sun is composed mainly of hydrogen and helium with traces of many other elements including carbon, nitrogen, oxygen, neon, magnesium, silicon, sulfur, argon, calcium, iron, and nickel.

The Sun emits huge amounts of energy and mass. It radiates enough energy in 1 second (4×10^{33} erg/sec) to power 4 million cars for 1 billion years. Analysis of the solar spectrum reveals a wealth of information about the Sun and its physical properties. The "quiet sun" emission is the total background output, excluding the discrete localized surface sources (flares, etc.), which takes the form of the black body (thermal) radiation emitted from the randomly moving particles making up the hot solar gas. However, the apparent "brightness temperature" is a function of the type of electromagnetic radiation within the solar atmosphere. For example, at visible wavelengths the brightness temperature is about 6000 K, but at both wavelength extremes (x-ray and radio) the Sun's brightness temperature is over a million degrees. As we will soon see, these temperature differences reflect the fact that solar atmosphere has very distinct temperature regimes which radiate at their own effective temperature. The solar atmosphere has a minimum temperature of 4600 K near the Sun's surface (photosphere-chromosphere boundary) and reaches a maximum temperature of a few million degrees at heights of nearly a million kilometers above the solar surface.

The total amount of energy per second, from all wavelengths, received at the top of Earth's atmosphere is called the solar constant. The presently accepted value of the solar constant is 1370 watts per meter squared at a Sun-Earth separation of 1 astronomical unit (1 AU is about 93 million miles). The major contribution to the solar constant is from the infrared (52%) and visible (41%) portions of the solar spectrum. The near ultraviolet (<7%), extreme ultraviolet (0.1%) and radio waves (<0.1%) round out the other principal energy sources. The major radiation sources (infrared, visible, and near ultraviolet) seem to be constant to within 1%. Fortunately the radio and x-ray wavelengths, which show huge fluctuations, make only a minor contribution to the solar constant.

It is only natural to ask if the solar constant is really constant? Over the last decade, spacecraft observations showed the solar constant varied only at a level on the order of 0.1%; over the past few years it appears the solar constant is decreasing by about 0.018%/year. Even though small, this change in the solar constant appears to have a secular component which is generally in phase with the solar cycle. We used the phrase "appears to have a secular component" because the details on this secular variation are based on just 10 years of satellite data, and therefore should be viewed with a bit of caution. For example, observational evidence points to the solar cycle changes in the solar faculae structure (see section 2.3), and the resulting small increase in disk brightness, as the source of this variability. However, the scientific community is still arguing whether such nearly constant behavior is a short term quirk or truly a characteristic of our Sun.

Some scientists argue that solar constant variations larger than 0.1% may occur on time cycles longer than the 11 year solar cycle. In fact, models of the Earth's atmosphere show that just a 0.5 to 1.0% change in the solar constant can account for a 1 to 1.5°C temperature change on Earth. Therefore, small changes in the solar constant could account for many of the variations in the Earth's climatic history. Furthermore, these scientists argue that astronomical evidence shows that stars similar to our Sun normally undergo significant brightness variations. As such, they argue, the solar constant is unusual by its very consistency. Regardless of what the future may hold, we will assume for the purposes of this text that the total solar brightness can be treated as a constant.

The Sun also emits a steady stream of particles (protons and electrons), known as the solar wind. The solar wind carries about one billion kilograms of mass away from the

Sun every second. Even at this rate, it would take over 10^{12} years to decrease the mass of the sun by just 10%.

2.1 Stellar Structure

The nuclear fusion of hydrogen into helium is thought to be the source of the Sun's energy. Fusion is confined to the core of the Sun (innermost 20%) where nearly 600 million metric tons of protons are consumed per second to produce helium nuclei (alpha particles). Eventually, the core hydrogen will be exhausted and the Sun will begin a sequence of expansion and then contraction, before it dies as a burnt out superdense hulk about the size of the Earth with nearly its present mass. Sufficient fuel remains in the core to power the Sun for at least another 5 billion years.

The tremendous energy generated in the core of the Sun is transported to the surface by three mechanisms: conduction, radiation, and convection. Conduction is present to a small degree throughout the Sun, but it does not appear to be dominant anywhere. Radiation is energy transport by photons and it dominates in the high density core region. The convection process dominates in the stellar envelope region which surrounds the core (see Figure 2.1). The envelope comprises the majority of the solar volume and the top of the envelope is referred to as the surface of the Sun. Due to the convective nature of the envelope region, the Sun's surface appears as a boiling, granular layer which resembles the surface of an orange (see Figure 2.2).

The individual convective cells (granules) are typically about 1000 km in diameter and have average lifetimes of about 8 minutes. Each granule consists of a vertically rising central region with velocities less than 1 km/sec, which then

spreads horizontally outward at the top of the cell at speeds of about $\frac{1}{2}$ km/sec. Cooler gas descends at the edge of each cell in the darker (cooler) intergranular region which has a spacing comparable to the granular diameter. Mixing length theory assumes that the vertical scale of a convection cell is comparable with the horizontal scale, and thus granules probably extend downward at least 1000 km. This penetrative depth is small compared to the estimated 200,000 km deep outer convecting layer of the Sun's envelope.

On a much larger scale there are systematic horizontal patterns with dimensions of about 30,000 km that persist for more than an hour. These supergranular cells have a well-established central outflow of about 0.5 km/sec and a poorly defined downflow. It is generally accepted that supergranules are convection-like cells with penetrative depths of about 10,000 to 15,000 km. Supergranules contain hundreds of individual granules.

There is also some evidence to support an intermediate 7000 km mesoscale granules. There is also some suggestion of giant convection cells which extend over the 200,000 km depth of the convective zone. These giant cells have been used to explain the short term variation in the solar equatorial rotation rate. Since the Sun is made up entirely of gas, it doesn't rotate as a rigid body. The solar equator rotates at about 2.9×10^{-6} radians per second and decreases with solar latitude to about 2.0×10^{-6} radians per sec near the poles. There is substantial evidence showing that the solar rotation rate has increased by about 2% in the equatorial region and about 4% at higher latitudes, over the last two solar cycles. Even though there are no direct measurements of giant cell velocity fields, most global circulation models assume that giant cells exist.

2.2 Solar Atmosphere

Above the solar surface is an extensive atmosphere which extends out beyond 10 solar radii. The solar atmosphere consists of three regions: photosphere, chromosphere, and corona. The photosphere (or "light sphere") lies at the top of the convective zone and its most distinguishing feature is the polygonal, bright granules discussed in the previous section. Another important photospheric feature is sunspots which will be discussed in a later section of this chapter. Some authors call the photosphere the Sun's surface, while others (including the author) like to fold the photosphere into the solar atmosphere. The difference in these two interpretations is at most cosmetic. The gaseous Sun has no hard surface like the Earth, and the photosphere is really the transition or boundary between the denser solar envelope and the tenuous solar chromosphere. All agree that the photosphere is opaque which means that it is the deepest we can "see" into the Sun, and it is by extrapolating observations of the visible solar features that we have some understanding of the solar interior.

THE SUN: A MODEL

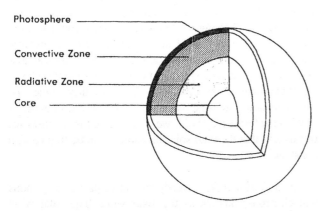

Figure 2.1 Cutaway showing the interior structure of the Sun.

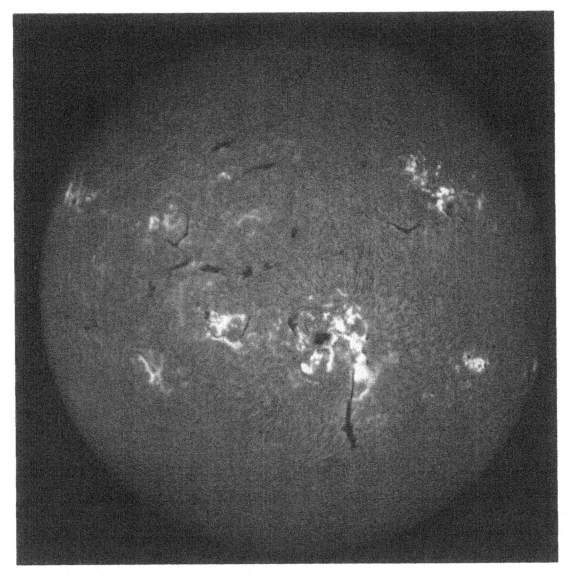

Figure 2.2 Hydrogen-alpha photograph of the solar chromosphere showing plage areas (bright regions), filaments, and granules *(courtesy Space Environment Laboratory).*

The photosphere is not equally bright over the solar disk. The edge, or limb, looks darker than the center of the solar disk because the photosphere can be observed down to a depth of a few hundred kilometers before it becomes completely opaque. Therefore, an observer can see the deepest, hottest, and brightest layers of the photosphere at the center of the solar disk. However, looking toward the limb, the observer's line of sight cannot see as deep into the photosphere and so one only sees the cooler (dimmer) upper regions of the photosphere. Studies of this limb darkening effect show that the photospheric temperature ranges from a maximum of 6000 K at the deepest portion of the photosphere to a minimum of 4300 K near the photosphere-chromosphere boundary. At these temperatures, most of the radiated photospheric energy is in the visible wavelengths. These studies also indicate that the mean photospheric gas

density is about 10^{14} particles per cm^3 (compared to 10^{19} particles per cm^3 for the Earth at sea level) with a resulting gas pressure of about 0.01 Earth atmosphere. This pressure is the product of the mass of the tenuous solar atmospheric gas lying above the photosphere combined with the Sun's large gravitational field.

Above the photosphere is the atmospheric layer called the chromosphere or "color" sphere which is named after the brilliant red flash seen from this layer during a solar eclipse. The chromospheric gas is less dense and hotter than the lower photosphere. The base of the chromosphere is defined as the height of the temperature minimum in the solar atmosphere, 4300 K. Although thin (approximately 3000 km), the chromospheric temperature increases rapidly with height, reaching a temperature of nearly 25,000 K at the boundary with the transition region. Within the transition region the temperature quickly increases to nearly a million degrees before it merges with the overlying corona. Since neither the chromosphere nor the corona are in thermodynamic equilibrium, these temperatures are called kinetic (as in kinetic energy) temperatures rather than the more familiar sensible temperatures that we "feel" with our bodies. The particle density decreases steadily through the chromosphere, reaching a value near 10^{10} particles per cm^3 at the top of the layer.

How can the cooler photosphere heat the chromosphere to nearly a million degrees? A popular theory is that energy is transferred to the chromosphere primarily by a mechanical process. Low frequency sound waves generated in the turbulent convective zone move upward into the tenuous chromosphere and become shock waves that induce energetic collisions which heat the chromospheric gas. Other possible heating mechanisms are by magnetohydrodynamical (MHD) waves, and by ohmic heating due to currents flowing in the solar atmosphere. However, to date there is no well-accepted theory of chromospheric heating. Similar mechanisms are used to explain the transition and coronal region heating, but once again there is no one well-accepted heating mechanism.

Above the chromosphere is the very hot, tenuous outer layer of the solar atmosphere called the corona (or "crown" layer). Coronal kinetic temperatures exceed two million degrees, and particle densities are below 10^5 particles per cm^3. The corona's low density accounts for its low luminosity and it is only visible when the bright disk of the Sun is occulted during an eclipse or artificially by a coronagraph.

Due to the high coronal temperatures, the corona is fully ionized, and therefore, it is affected by the solar magnetic field. Closed magnetic field lines (solar field lines which do not extend much beyond the solar atmosphere) will trap the coronal plasma and thereby keep the particle density relatively high. Open field lines (solar field lines extending deep into interplanetary space) allow the coronal plasma to escape and these low density coronal areas are called coronal holes. We will discuss coronal holes in more detail in Chapter 3, and we will discuss the current theory that coronal holes are the source of high speed solar wind streams.

Since the coronal temperature increases with height, the limb appears brighter (in x-ray wavelengths) than the center because within the very low density coronal gas there are more emitting sources seen along the longer line-of-sight paths from the limb region. Coronal limb brightening is due to higher numbers of emitters whereas photospheric limb darkening was a result of high absorption which limited emission sources to the cooler portions of the photosphere.

2.3 Active Sun

Since the energy from the core of the Sun is released at a steady rate, one might expect a spatially uniform and tempo-

rally steady release of radiation from the solar surface. However, over short periods of time and in certain locations, the solar intensity can fluctuate rapidly. It is thought that a major factor causing these fluctuations is the distortion of the Sun's large-scale magnetic field (1–3 gauss at the surface) due to differential rotation. As we stated earlier, being a gaseous body, the Sun need not rotate as a rigid body. In fact, the solar equator rotates faster than regions near the poles: 24.9 days at the equator compared to approximately 31.5 days near the poles. We believe that this differential rotation causes the Sun's main magnetic field to become twisted and concentrated into specific regions, and then this stored magnetic energy is released as plasma kinetic energy when the field relaxes back to its initial state. Such regions are called "solar active regions" or simply "active regions." The most common indications of locally enhanced magnetic fields are facula, plages, spicules, prominences, sunspots, and flares (see Figure 2.2).

Solar observers have frequently observed vast chromospheric regions of enhanced solar emission due to a higher temperature and density of the chromospheric gas in these areas. Early French observers compared the appearance of these regions to the white sand of a beach and named them plage, the French term for beach. A plage contains a magnetic field of 200–500 gauss, compared to 1 to 3 gauss for the undisturbed Sun. When a plage forms, it is relatively small and intense. As it grows in size, it initially maintains its intensity, but about midway in its life, the plage decreases in intensity while continuing to grow in size. A large plage covers a few percent of the solar surface; by comparison, a disk the size of the Earth would cover less than 0.01% of Sun's surface. The lifetime of a plage may be several days, weeks, or even months.

Sometimes, long-lived, enhanced white light regions appear in the photosphere. These regions, called facula ("little torches"), usually occur in conjunction with chromospheric plages. Facula appear before the emergence of sunspots and persist long after the spots have faded away. Sometimes plage regions are referred to as "chromospheric facula."

Within plage regions, an intricate fine structure appears whose collective appearance is referred to as filigree, which consists of roundish dots and slightly elongated segments with dimensions on the order of a few hundred kilometers. The filigree elements tend to lie between granules and they change in shape and sharpness over a time scale comparable to the lifetime of granules.

The chromosphere also contains many small jet-like spikes of gas rising vertically at speeds of up to 30 km/sec. These features, called spicules, occur at the edges of supergranule cells, and when viewed near the limb of the Sun so many are seen in projection that they resemble a forest or possibly a prairie fire. Individual spicules have lifetimes of about 10 minutes. Spicules reach greater heights over the poles, and giant spicules (or macrospicules) have been found in these regions. Typical spicules have emission temperatures of 10,000 to 20,000 K and are about 1000 km broad and rise to heights of about 10,000 km. Macrospicules exceed a typical spicule by a factor of four or five in all parameters: temperature, width, height, and lifetime.

Another chromospheric feature are fibrils which are short-lived (10 to 20 min) horizontal strands of gas on the order of 10,000 km long and 1000 to 2000 km wide. The pattern of fibrils follows the local magnetic fields. For example, fibrils show up as dark absorption features following the lines of force between sunspots of opposite polarity, giving a pattern similar to that obtained when iron filings are scattered on a sheet of paper placed over a bar magnet.

Most chromospheric observations are made at the strong hydrogen-alpha (6563 Å) emission line. At this H-alpha wavelength, we frequently see prominences which are stable structures, millions of kilometers long on the solar limb. Superficially, prominences appear to be material ejected upward away from the Sun, but motion pictures show that the coronal gas condenses at the top of the arch and appears to move downward along closed magnetic field lines. We also see dark, stable, threadlike features against the disk of the Sun which are called filaments. Prominences and filaments are different names for the same phenomena. When seen on the disk the suspended gas absorbs more photospheric energy than it gives off along the line of sight, and thus it appears dark. When seen on the limb, the condensed gas emits more energy than the tenuous corona that surrounds it, so it appears bright. Prominences usually originate near regions of sunspot activity and lie on the boundary between regions of opposite magnetic polarity. In general, prominences may be divided into two basic categories: quiescent and active.

The long-lived, quiescent prominences appear as thin ropes of material which are usually less than 3000 kilometers in diameter and tens of thousands of kilometers long. They normally form within mature (or decaying) plage areas. As the prominence ages, it moves out of the plage region and slowly migrates toward the pole. Quiescent prominences may be visible for several solar rotations during which time they may undergo changes in shape and location ranging from slight perturbations (knots) to the complete eruptive destruction of the filament.

Figure 2.3 is a dramatic Skylab photograph of an erupting prominence using the HE 304 Å line. In the picture, the prominence has already moved nearly one solar radius above the solar surface. As the ejecta from an erupting prominence (or filament) moves outward, it carries a substantial amount of mass into the corona and eventually into the interplanetary medium. The resulting coronal mass ejection, or CME, can produce an interplanetary shock which, in turn, may induce a geomagnetic disturbance if the shock intersects the Earth's magnetosphere.

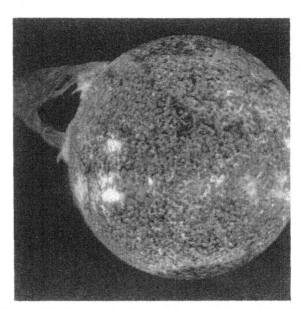

Figure 2.3 Photograph of an erupting prominence taken from Skylab using HE 304 Å filter *(courtesy Space Environment Services Center)*.

Erupting filaments can produce at least a factor of ten increase in the corona's background x-ray flux (sometimes the increase can be as much as 10^3). As a matter of comparison, a flare can generate up to a 10^5 enhancement in the x-ray solar flux. The enhancement from an erupting filament appears to start within tens of minutes after the eruption begins and can last for 12 hours.

As of this writing, there is no day-to-day capability to image erupting filaments because a soft x-ray imager needs to be in orbit above the Earth's atmosphere. Presently, the GOES weather satellite has a soft x-ray detector which integrates measurements over the whole solar disk. As such, this detector cannot pinpoint the location of solar x-ray events. Instead, filament eruptions are located after the fact by using the disappearance of a filament between one daily ground based H-alpha patrol photograph and the next. The H-alpha patrol photographs are part of the normal procedures of the U.S. space forecast centers. Plans are underway to fly a solar x-ray imager on the next generation of GOES satellites.

Active prominences include surges, sprays, and loops. Surges are radial projections of photospheric plasma into the corona at speeds of 100 to 200 kilometers per second. Sprays are fragmented pieces of lower atmospheric material which are ejected during energetic solar flares. A single loop prominence may start as a surge, but during the next several hours, a multiloop system may develop and attain heights of 50,000 km. In a loop system, the individual loops do not expand but rather fade and are replaced by higher ones. A loop prominence is usually evidence that a very energetic solar flare has occurred.

The basic magnetic connections between active regions, which appear as loops in the corona at x-ray wavelengths, seem to persist for the duration of those active regions, but individual coronal loops survive for much shorter times (usually less than a day). In contrast to chromospheric loop prominences, coronal loops represent quiescent coronal conditions. In the recent past, coronal models considered the role of magnetic fields as a means of modifying the otherwise average coronal temperature and density. Observationally, the corona's quiet "appearance" is a consequence of averaging over sufficiently large spatial and temporal scales. Recent observations show the corona is constantly active and restless, and that small, short-term processes are important. Today, more and more models consider the corona as a collection of magnetic loops with effectively nothing between them. The magnetic fields constitute the basic elements of the corona, and without the magnetic fields the nature (if not the very existence) of the corona would be much different than what we observe today.

Figure 2.4 is a soft x-ray (3–54 Å) photograph of the Sun taken by Skylab astronauts on May 28, 1973. The dark areas on the photograph are called coronal holes which have a magnetic geometry quite different than the loop structures characterizing the active regions. In a coronal hole, the magnetic field opens out into interplanetary space whereas the magnetic field in a coronal loop is rooted at both ends on the solar surface. The open magnetic flux tubes, making up the coronal holes, provide an easy avenue of escape for coronal plasma and the coronal holes are thought to be the source of the observed high speed solar wind streams (see section 3.5). The coronal hole in the top left part of the picture is over the solar north pole. The dark, horizontal line along the bottom of the image is called a filament cavity; as the name implies, the cavity is the coronal artifact of a chromospheric filament which lies below it. The bright area to the center right of the photograph is due to the higher density coronal plasma trapped by the magnetic loop structures associated with a large surface active region (region number AR 114) and a nearby smaller active region (AR 113). The bright area to the center left was above active regions AR 115 and 11, and the bright region on the west limb (right side of picture) was above AR 103 and 105. The bright regions on the west and east limb reveal some of the structure associated with the coronal loops. The point-like brightenings on the picture are called bright points and are associated with newly emerging, intense magnetic fields from the solar surface.

Unlike coronal holes, coronal mass ejections are produced by some type of energetic process in the solar atmosphere. Two obvious sources of these CMEs are flares and erupting filaments. There is strong observational evidence showing that all long duration (6 hours or longer) x-ray events, flares, or erupting filaments produce CMEs. However, other CMEs seem to originate from processes entirely within the corona, but identifying the precise physical mechanisms behind these CMEs is still a hotly debated issue.

CME occurrence varies with solar cycle. Spacecraft data indicates that during solar maximum CMEs occur at least once a day; however, during solar minimum, the CME rate drops to about once every 4 days. The average ejected mass appears to be about 4×10^{15} grams and the average kinetic energy is about 3.5×10^{30} ergs, with a range of about 100 for both parameters. CME's move away from the Sun with speeds as low as 14 km/sec and as fast as 1,800 km/sec with average speeds ranging between 340 to 472 km/sec; the range of values is due to differences in the satellite sensors used to collect historical data sets.

The faster CME's, at superalfvenic speeds (see Chapter 3) above 400 km/sec, are ideal candidates for initiating interplanetary shocks. In fact, satellite observations confirm this prediction. Coronagraph observations of CMEs near the Sun, followed by observations of interplanetary shocks near 0.5 AU, using zodical light photometers, fit the pattern predicted for simple shock expansion in the interplanetary medium.

2.4 Solar Flares—Observations

A solar flare is a highly concentrated explosive release of energy within the solar atmosphere which appears as a

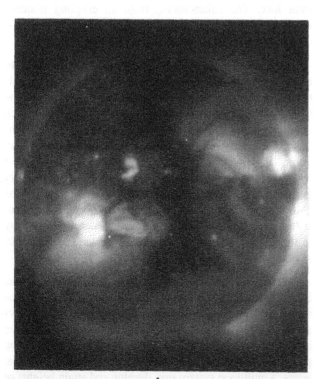

Figure 2.4 Soft x-ray (3–54 Å) Skylab photograph of the solar corona. The dark areas are called coronal holes which have an open magnetic field geometry allowing coronal plasma to easily escape *(courtesy Space Environment Services Center)*.

sudden, short-lived brightening of a localized area in the chromosphere. Solar flares are differentiated according to their total energy released because, ultimately, the total energy emitted is the deciding factor in the severity of a flare's effects on the near Earth environment. The radiation from a solar flare extends from radio to x-ray frequencies.

Flares are classified according to their size and intensity. The size or "importance" classification is based on flare area (in terms of the fraction [in millionths] of the solar disk) according to the following table:

Importance	Corrected Area (in millionths of the solar disk)
0 (subflare)	less than 200
1	200–499
2	500–1199
3	1200–2400
4	greater than 2400

The corrected area means that flare size is adjusted for solar curvature. A flare the size of the Earth would only be importance 0.

The brilliance of a flare is measured in two frequency bands: optical and x-ray. The most common optical flare intensity or "brilliance" classification is based on the doppler shift of the H-alpha line. This doppler shift is a measure of the ejected gas particle velocity and is used by observers to make a subjective estimate of flare intensity. Using this system, flares are classified as faint (F), normal (N), or brilliant (B). Unfortunately, the observed optical intensity is strongly dependent on seeing conditions, and only a slight amount of terrestrial atmospheric pollution can drastically alter the measured intensity. Moreover, the flare intensity is based on the brightest element of a flare. A flare is reported as "brilliant" if any portion reaches this level, even though most of the flare may be "faint."

While the visible (optical) emission from a flare increases by, at most, a few percent, the x-ray emission may be enhanced by as much as four orders of magnitude. Presently, x-ray sensors are located on some geosynchronous satellites, and therefore terrestrial atmospheric attenuation is not a problem. Measurements are made in the 1 to 8 Å (soft) x-ray band and the 1/2 to 4 Å (hard) x-ray band (1 Å $= 10^{-10}$ m). X-ray flares are classified according to the peak energy flux of the flare. For soft x-rays, the classification system is:

Classification	X-Ray Flux (ergs/cm²-sec)
C	10^{-3}
M	10^{-2}
X	10^{-1}

The above categories are broken down into nine subcategories based on the first digit of the actual peak flux. For example, a peak flux of 5.7×10^{-2} ergs/cm²-sec is reported as a M5 soft x-ray flare.

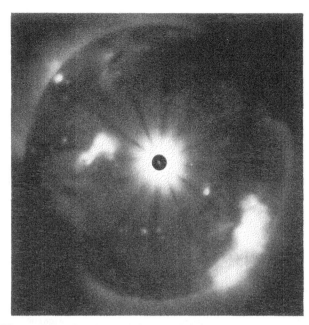

Figure 2.5 Skylab x-ray photograph of a solar flare; center of flaring region removed (dark disk) because flare brightness oversaturated the photographic plate *(courtesy Space Environment Services Center)*.

Figure 2.5 shows a Skylab x-ray photograph of a solar flare. The center of the flaring region has been removed and appears as a dark disk because the flare oversaturated the photographic plate. The ray-like features are also artifacts of the flare brightness oversaturating the camera.

The position of a flare on the visible disk is one of the factors taken into account when forecasting whether or not a flare will cause a disturbance at Earth. Figure 2.6 shows that flares located near the west limb of the Sun have a higher chance of interacting with the Earth than east-limb flares. Historically, flares occurring near 50° west longitude are most likely to generate terrestrial disruptions.

The total energy released during a flare may range from 10^{21} to 10^{25} joules (1 joule $= 10^7$ ergs) integrated over the three phases of a flare: precursor (or preflare), flash, and main (or gradual) phase. The precursor phase can last from minutes to hours, and is characterized with a slight enhancement of soft x-ray emissions and the emergence of magnetic flux from the photosphere. The flash phase begins with an increase in optical and x-ray emissions by at least 50% above background in one to five minutes. Within the flash phase there may be impulsive bursts of microwaves and x-rays. The spatial extent of the flare continues to expand until shortly after the flare reaches maximum intensity. The main phase can last for hours and is characterized by the slow decay to preflare levels. However, additional bursts may occur, although they rarely reach the flash intensity and they tend to be extended in time compared with the impulsive bursts. The

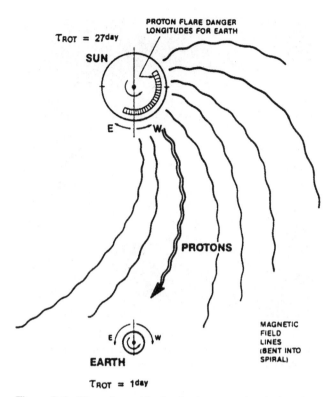

Figure 2.6 Flares located in the Sun's western hemisphere have the best chance of interacting with the Earth; flares near 50° west longitude have a history of generating terrestrial disruptions (*after Wagner, 1987*).

mean duration of an optical flare is loosely related to a flare's magnitude:

Importance	Average Duration	Percent of All Flares
0	17 minutes	75
1	32 minutes	19
2	69 minutes	5
3 or 4	more than 2 hours	less than 1

Solar proton events make up another important class of flares. These very large flares eject extremely energetic protons which arrive at Earth in 30 to 90 minutes after flare initiation. Remember, the Sun is only slightly more than 8 light-minutes from the Earth and, therefore, solar proton flares accelerate particles to a substantial fraction of light speed. As shown in Figure 2.7, proton events can last from 2 to 6 hours, with the lower energy tail of the proton particle distribution taking the longest time to arrive at Earth. There is also considerable observational evidence showing that CMEs may be linked to these very energetic flares.

Historically, proton events were detected at Earth after the energetic solar protons had severely disrupted the high latitude ionosphere causing widespread absorption of radio signals along transpolar cap circuits. Today, solar proton events are first detected as a large, brilliant class flare with a unique radio emission spectra as measured by Earth based solar radio telescopes. However, the actual arrival of the energetic protons is detected by a pair of high energy particle sensors aboard the GOES satellites; the sensors have thresholds at 10 MeV and 30 MeV. An actual proton event is verified when the 10 Mev proton flux exceeds $10 \text{ cm}^{-2}\text{s}^{-1}\text{sr}^{-1}$ which is approximately four times the background flux at

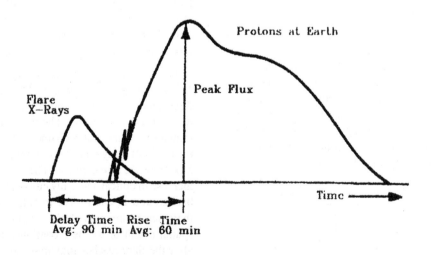

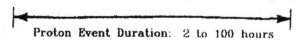

Figure 2.7   The time sequence of events for proton flares (*after Wagner, 1987*).

Proton

Number of Proton Events by Month

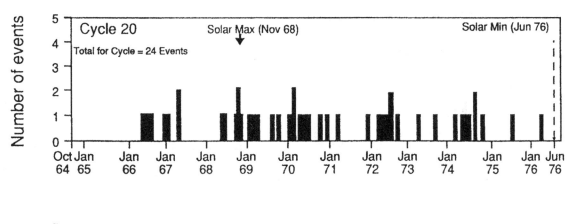

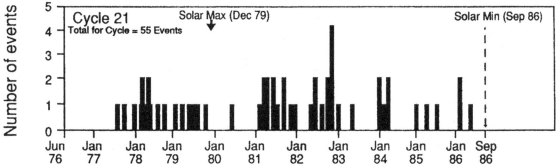

Figure 2.8 Frequency of occurrence of solar proton flare events during solar cycles 20 and 21 *(courtesy Space Environment Services Center).*

geostationary altitudes. Figure 2.8 shows the frequency of solar proton events over the last two solar cycles.

Another way to detect the most energetic solar proton events is to measure the excess number of neutrons created as secondary products after the protons collide with and shatter atmospheric atoms. Normally, there is a nearly steady background flux of such neutrons generated by galactic cosmic rays. Detectors at Thule and McMurdo Station are designed to measure neutrons created by cosmic rays with energies greater than 0.4 GeV. During a large proton event, the neutron flux can exceed the background value by an order of magnitude or more, and the resulting "neutron shower" is called a ground level event (GLE). Figure 2.9 shows the frequency of GLEs during a solar cycle. Although infrequent, GLEs can occur in rapid succession. For example, from July to November 1989, seven GLEs were recorded.

2.5 Flare Theory and Dynamics

Various models of solar flares have been proposed, but none explains all flare observations. It is currently believed that the

solar flare energy is probably stored in the twisted and kinked magnetic field lines above active regions in upper portions of the solar atmosphere. There are basically three possible flare schemes. First, all the magnetic energy is stored in situ above the photosphere in some equilibrium or slowly evolving steady state and is then released. Second, an externally driven mechanism, such as a double layer (see section 8.6), causes the total stored energy to be released from some given volume. Finally an externally driven mechanism releases some fraction of the stored energy, which in turn propagates outward, expanding the available energy supply by destabilizing other atmospheric magnetic features, such as prominences. The biggest problem with any flare model is the release of sufficient energy needed to accelerate an estimated 10^{33} to 10^{36} electrons per second, with each electron having about 25 keV of energy. Relativistic electrons are produced only during the infrequent large flares that have durations of many hours to days.

It is generally believed that the flash phase is associated with the primary energy release mechanism which, in turn, is responsible for the impulsive bursts. Whereas, the extended

GLE

Number of Ground Level Events by Month

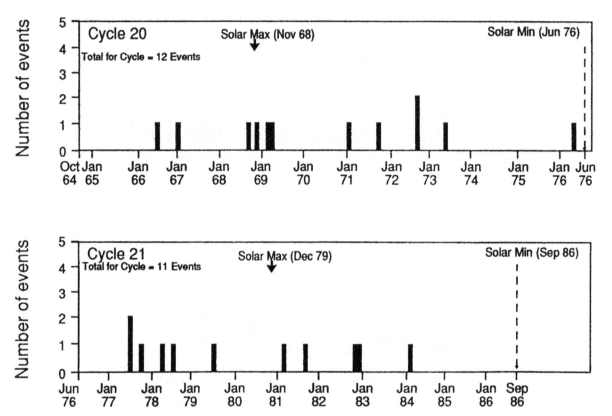

Figure 2.9   Frequency of occurrence of ground level events (GLEs) during a typical solar cycle *(courtesy Space Environment Services Center)*.

bursts of the main phase are believed to be associated with a secondary energy release due to shocks, excited by the primary energy mechanism, propagating out into the corona. Observations show that flare kernels (patches of bright emission) brighten close to the emerging magnetic flux regions found within an active region. However, it is not known if the emerging flux acts as the trigger or is just part of a complex process which involves the regions both below and above the photosphere.

Energetically, a low β plasma configuration is preferable for solar flare mechanisms because only a small fraction of the magnetic field needs to be dissipated in order to heat the surrounding plasma. Remember that β is the ratio of plasma pressure over the magnetic pressure, and that a low β plasma implies that the magnetic fields dominate the dynamics. For example, suppose beta is about 10^{-3}. If 10% of the magnetic energy is transferred by some process into heating the nearby plasma, then the plasma temperature can be increased by a factor of 100 and β will increase to 0.1, assuming the plasma density doesn't change. Thus, even after the plasma has

undergone significant heating, the magnetic field still dominates and the plasma remains constrained as is observed in small flare loops.

Each flare has its own distribution of energy by frequency. If flare A is twice as intense in x-ray frequencies as is flare B, then flare A will not necessarily be twice as intense as flare B in optical frequencies. At any given frequency, the typical flare shows a near-exponential rise in intensity during the explosive initial (flash) phase and then a slow decay to pre-flare levels. Flares which release a greater portion of the energy at higher frequencies ("hard" spectrum flares), cause the greatest disruption of the near Earth environment. Almost immediately following these flares, the enhanced x-ray flux will intensify (harden) the Earth's ionosphere which, in turn, disrupts HF communications and long-range radars (see Chapter 9). A few days following a flare, ejected solar material may interact with the Earth's magnetic field and again may disrupt communications, long-range radars, and satellite operations.

The frequency of optical solar flares during cycles 20–21 is shown in Figure 2.10; the data represents smoothed Zurich

Optical Flares
Number of Optical Flares by Month

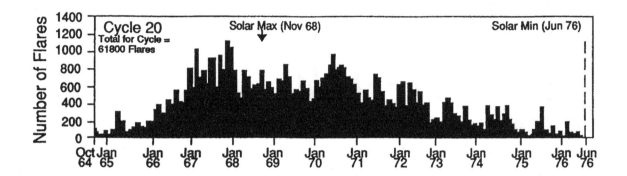

Total as of equivalent time cycle 22 (Apr 67) = 9683 Flares

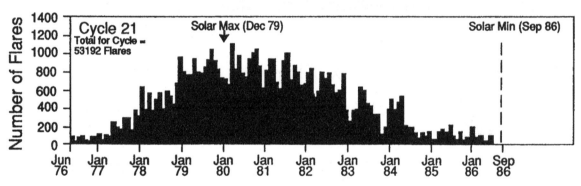

Figure 2.10 Frequency of occurrence of optical flares during solar cycles 20 and 21 *(courtesy Space Environment Services Center)*.

sunspot numbers (see Section 2.7). Figure 2.11 shows the M and X class x-ray flares as measured by the GOES x-ray detector. Remember this instrument cannot differentiate between flares, eruptive prominences, or other energetic coronal events which generate soft x-rays. Therefore, a few of the recorded M-class events may not be due to solar flares. Notice the occurrence of solar proton events which appear as peaks on either side of sunspot maximum. This phenomena is common to all solar cycles and may be associated with the increasing flare rate near sunspot maximum. The increasing number of smaller flares around sunspot maximum may relieve enough magnetic stress to significantly reduce the chances for very large flares.

2.6 Sunspots

Sunspots are the most commonly reported solar activity feature. The term *umbra* is applied to the dark core of the spot and *penumbra* to the gray boundary region surrounding

the umbra. Sunspots are regions of very strong magnetic fields (typically several thousand gauss) and it is thought this intense field confines the gas within the spot and reduces the interaction between the spot and the surrounding gas. In part, this lack of interaction allows the gas in the sunspot to cool relative to the surrounding photospheric plasma by approximately 1000 K, thus creating its dark appearance. Sunspots are located in the photosphere under plage regions, but note: not all plage regions have sunspots beneath them. Less than half of all sunspots develop a penumbra, and today there is some evidence that the penumbra is probably more dependent on sunspot magnetic field geometry than on spot size or intensity. Dark, mature penumbras are often associated with older sunspot groups.

Penumbras are typically 2.5 times the diameter of the umbra and umbras usually range from 300 km to 2500 km in diameter; the largest umbra observed with no associated penumbra was 11,000 km in diameter. The penumbra is the site of the most pronounced photospheric velocity feature,

M-Flares
Number of M Class X-ray Flares by Month

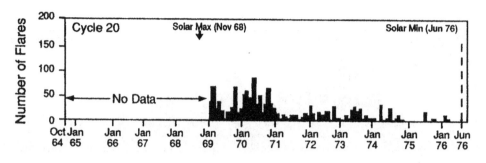

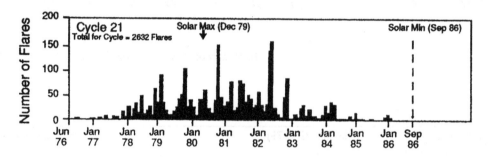

X-Flares
Number of X Class X-ray Flares by Month

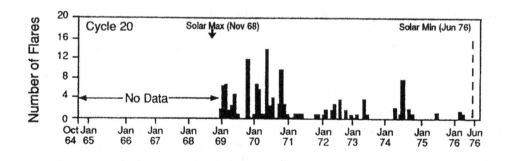

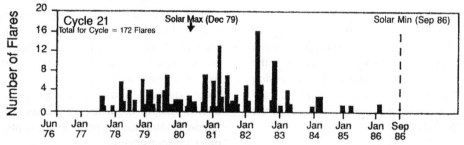

Figure 2.11   Frequency of occurrence of M and X class x-ray solar flares during solar cycles 20 and 21 *(courtesy Space Environment Services Center)*.

the "Evershed effect." The umbra itself is remarkably free of visible motions but the penumbra has observable horizontal velocities out of the sunspot, toward the surrounding photosphere, with average velocities approaching a few kilometers per second. It is this rapid horizontal acceleration within the penumbra that we call the Evershed effect. Observations indicate that the outward horizontal velocity flow decreases with altitude, and in the chromosphere the flow reverses direction and moves toward the center of the sunspot. It appears that there is a circulation of solar atmospheric gas that flows out through the penumbra, rises up vertically, and then flows toward and back down to the umbra.

The distinction between pores and sunspots has led to some confusion because some people have defined pores as any sunspot lacking a penumbra. However, more than half of the spots never develop penumbra. Among observers, pores have been traditionally defined as a darkening in the photosphere similar in intensity to the intergranular spaces, with a diameter up to 2500 km, and a magnetic field strength in excess of 1500 gauss. The lifetime of a pore is a function of the region where it forms. In a quiet photosphere, pores have a lifetime of 10 to 15 minutes but near a sunspot group a pore can persist up to an hour. Also, some photospheric regions may form pores repeatedly in the same area for more than a day. A pore may be the initial stage in forming a sunspot with the transition being one of continued darkening and growth rather than a coalescing of two or more pores.

A large percentage of solar regions develop more than one sunspot and this collective unit is termed a sunspot group. A sunspot group is oriented primarily east-west where the directions east and west are with respect to an observer standing on the Earth. At sunrise, the west limb of the Sun will rise first, and at midday the Sun's west limb is nearest to the Earth's western horizon. The western-most sunspot in a group is known as the leader (or preceding) sunspot, and the more easterly spots are called follower (or trailer) spots.

The leader spot is of special interest in studying the group. Generally, it is the first spot to form and the last spot to disappear. Typically, it is also the first spot in the group to form a penumbra and last to lose it, and the leader spot is the largest spot throughout the life of the group. Also, the sunspot groups in a given solar hemisphere tend to have leaders of the same magnetic polarity, and the leaders in the northern hemisphere usually have the same magnetic polarity as the trailer spots in the southern hemisphere. Furthermore, the leader spot polarity usually matches the background magnetic field polarity of its resident hemisphere. Periodically (approximately every 11 years) the hemispherical magnetic field polarity reserves. If the leader spots in the northern hemisphere have north magnetic polarity during one 11 year epoch, then they will have south magnetic polarity during the next 11 year epoch. This periodic reversal of the solar magnetic field is part of the process called the

solar cycle. Another important aspect of the solar cycle is the periodic variation of the number of sunspots.

Sunspot groups are often classified according to the Zurich Sunspot System which uses the letters A through F, plus H to describe sunspot group complexity. An A class sunspot group is a simple unipolar spot, whereas a B class group contains bipolar spots without penumbra. A class C group has bi-polar spots with a penumbra on the leader spot. Class D, E, and F groups show penumbra on spots of both polarity, and each class is differentiated by increasing complexity of the sunspot group and the size of the group in heliographic degrees. The H class is assigned to a sunspot group in the final stages of evolution when all that remains is a leader spot with a well-defined penumbra; the last stage occurs when the leader spot decays to class A. Many groups decay to class H after reaching C or D, and half of all groups never progress beyond A and B.

Currently, the U.S. space environmental forecast centers use the three-parameter McIntosh sunspot classification system. The McIntosh system starts with the Zurich sunspot number and adds two other parameters to describe sunspot complexity and stability. The second parameter describes the penumbra: x for no penumbra, r for rudimentary, s for small symmetric, a for small asymmetric, h for large symmetric, and k for large asymmetric. Symmetric sunspots are usually longer lived than asymmetric spots. The third parameter categorizes sunspot groups according to the relative crowding of spots in the group: x for a single spot group, o for an open group, i for an intermediate group, and c for a compact group. As you might imagine, an open group has a weak magnetic gradient between spots of opposite polarity, whereas compact groups have strong field gradients with the spots in close proximity to the neutral line.

The pattern of magnetic field lines between spots of opposite polarity in a bipolar group is easily recognized by the pattern of fibrils which appear in H-alpha (or K-line of ionized calcium) photographs. The highly ionized, hot solar atmospheric gas has an extremely high electrical conductivity which effectively traps the magnetic fields with the solar plasma. Therefore, in low β plasmas (as is the case near sunspots), the magnetic field controls the plasma phenomena. In their study of solar phenomena, solar astronomers have found that the action of the "neutral line" in sunspot groups is a good indicator of future solar activity. The neutral line (or line of inversion) separates regions of opposite polarity in a sunspot group and thus is the line where the vertical component of the magnetic field is zero. Unfortunately, for a given sunspot group, the neutral line is rarely reasonably straight, but instead is intertwined between the multitude of spots with varying polarities. For example, filaments are frequently found lying near neutral lines, and flares also tend to occur around these boundaries.

Figure 2.12 shows a daily, chromospheric neutral line map superimposed on a Skylab x-ray photograph taken on the

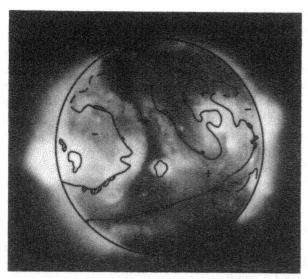

Figure 2.12 A chromospheric neutral line map is superimposed on a Skylab solar x-ray photograph taken on the same day; the plus/minus signs refer to the outward/inward orientation of the z component of the coronal magnetic field *(courtesy Space Environment Laboratory)*.

same day. On such a neutral line map a plus sign (+) refers to the outward direction of the z component of the solar magnetic field, and a minus sign designates areas where the solar magnetic field has a downward (negative z) component; hatched areas denote the location of filaments.

2.7 Solar Cycle

Heinrich Schwabe found in 1852 that the long-term yearly average sunspot number varied regularly with roughly a 10 year period. The time of largest average sunspot number was labeled solar (or sunspot) maximum, and that of lowest average number of sunspots was solar (or sunspot) minimum. More recent analyses have shown the average period length is about 11 years, with a spread of periods from 7 to 13 years. Each cycle is defined as beginning with solar minimum and lasting until the following solar minimum. The typical cycle takes four years to rise from minimum to maximum, and about seven years to fall back to minimum. Solar cycle 21, which began in June 1976, reached its maximum during December 1979. Cycle 22, which began in September 1986, reached its maximum in 1991.

There is some evidence that, at times, the sunspot cycle may completely disappear. For example, in 1893 E. W. Maunder of the Royal Greenwich Observatory concluded from a study of old solar records that there was a period of 70 years (1645 to 1715) when sunspot activity virtually vanished. More recent investigations seem to support the so-called "Maunder minimum," and imply that similar periods of inactivity reoccur on an irregular basis.

The most common index of solar activity is the Zurich sunspot number R. This number incorporates measurements of the number of sunspot groups and the number of individual spots. It also contains a correction factor which attempts to compensate for systematic differences in observations caused by variations in telescope size, atmospheric conditions, and the overenthusiasm of observers. Daily Zurich sunspot numbers are available back to 1749. The largest sunspot cycle on record is cycle 19 which peaked in 1957–58 with an R value near 200. The daily sunspot number varies more widely than does the long-term average. During solar maximum, daily sunspot numbers over 100 are normal and during solar minimum, several consecutive spotless days are possible.

The mean latitude of sunspot groups varies with the solar cycle. At the beginning of a new solar cycle (just after solar minimum), the average spot group will form near 40° solar latitude. As the cycle progresses, spot groups will form at successively lower latitudes until solar minimum, when groups form near the solar equator. Sunspots seldom occur poleward of 40° and never on the equator. The highest latitude group ever recorded was near 60° N solar latitude. The progression of regions of spot activity toward the equator during each solar cycle is known as Sporer's law.

A popular theory of this cyclic motion in latitude involves the production and relaxation of magnetic stress by differential rotation. The differential rotation of the Sun initially produces the greatest stress on the magnetic field near 40° latitude. This stress causes the field lines to erupt through the photosphere and become the seed for the development of sunspots. While the magnetic stress at 40° is being relaxed by solar flares, the stress continues to grow at lower solar latitudes as the solar cycle continues. Thus, the active solar latitude (the latitude of greatest magnetic stress) moves slowly equatorward.

As mentioned earlier, the magnetic polarity of the Sun's poles reverses every solar cycle. Early in the solar cycle, the polarity at one pole is positive (magnetic field lines pointing away from the Sun), the other negative (pointing toward the Sun). Near solar maximum there seems to be no dominant polarity at either pole, and a reversal of the Sun's magnetic polarity begins to become apparent within two years following solar maximum. During any solar cycle, the leader sunspots normally have the polarity that their hemisphere had at the beginning of the cycle, so at solar minimum the Sun may simultaneously have "old cycle" groups near the equator (with old cycle leader polarity) and "new cycle" groups near 40° (with new cycle leader polarity).

2.8 Summary

In summary, the Sun is the only star we can study closely. We believe the Sun makes its radiant energy by controlled thermonuclear fusion. In the superhot core, hydrogen nuclei are fused into helium nuclei, releasing a large amount of energy to be radiated into space. The mass of hydrogen still left in the Sun's core is sufficient to keep it going at the present rate for another 5 billion years. The 11 year sunspot cycle is tied to the frequency of disturbances on the Sun including flares, plages, and large sunspot groups. The change of solar activity with time can seriously affect today's sophisticated space systems. Both solar electromagnetic radiation and emitted particle fluxes are of concern (but for different reasons) as we shall see in the next chapter.

2.9   References

Athay, R. G. 1981. The Chromosphere and Transition Region, in *The Sun as a Star*, Stuart Jordan, ed. NASA SP-450, Chapter 4.

Beckers, J. M. 1981. Dynamics of the Solar Photosphere, in *The Sun as a Star*, Stuart Jordan, ed. NASA SP-450, Chapter 2.

Bieber, J. W., Everson, P., and Pomeranty, M. A. 1990. A Barrage of Relativistic Solar Particle Events, *EOS*, vol. 71, no. 33, p. 961.

Bostrom, C. O., Fisher, C. L., and Webb, P. 1987. USAF Scientific Advisory Board Report on Solar Flare Hazards to Man in Space, Pentagon, Va.

Brown, J. C., Smith, D. F., and Spicer, D. S. 1981. Solar Flare Observations and Their Interpretations, in *The Sun as a Star*, Stuart Jordan, ed. NASA SP-450, Chapter 7.

Cowen, R. 1991. Fortnight of Flares Dazzles Astronomers, *Science News*, vol. 139, no. 25, p. 139.

Dennis, B., and Canfield, R. 1988. Committee report on MAX '91—Flare Research at the Next Solar Maximum. NASA Goddard Space Flight center, Greenbelt, Md.

Eddy, J. A. 1979. *The New Sun: The Solar Results from Skylab*, NASA SP-402.

Gibson, E. G. 1973. *The Quiet Sun*. NASA SP-303.

Hargreaves, J. K. 1979. *The Upper Atmosphere and Solar-Terrestrial Relations*. Van Nostrand Reinhold, New York, Chapter 10.

Hirman, J. W., Hackman, G. R., Greer, M. I., and Smith, J. M. 1988. Solar and Geomagnetic Activity During Cycle 21 and Implications for Cycle 22, *EOS*, vol. 69, no. 42, p. 961.

Howard, R. J., Wagner, W. J., and Maltson, R. 1990. *Orbiting Solar Laboratory*, NASA NP-143.

Hundhausen, A. J. 1972. *Solar Wind and Coronal Expansion*. Springer-Verlag, New York.

Jensen, E. 1973. *Cosmical Geophysics*, A. Egeland, O. Holter, and A. Omholt, eds. Universitelsforlaget, Sweden, Chapter 2.

Jordan, S. D. 1981. Chromospheric Heating in *The Sun as a Star*, Stuart Jordan, ed. NASA SP-450, Chapter 12.

Kahler, S. 1987. Coronal Mass Ejections, *Reviews of Geophysics*, vol. 25, no. 3, p. 663.

Klecker, B., Cliver, E., Kahler, S., and Cane, H. 1991. Particle Acceleration in Solar Flares, *EOS*, vol. 71, no. 39, p. 1102.

McIntosh, P. S. 1979. Committee Chairman. Long-Term Solar Activity, in *Solar-Terrestrial Predictions Proceedings*, R. F. Donnelly, ed. NOAA/SEL, Boulder, Col., p. 246

McIntosh, P. S. 1981. The Birth and Evolution of Sunspots: Observations, in *The Physics of Sunspots*, L. E. Cram and J. H. Thomas, eds. Sacramento Peak National Observatory, Sunspot, N. M., p. 7.

McIntosh, P. S., Krieger, A. S., Nolte, J. T., and Vaiana, G. S. 1976. Association of X-Ray Arches with Chromospheric Neutral Lines, *Solar Physics*, 49, p. 57.

Moore, D. R. 1981. Penetrative Convection, in *The Sun as a Star*, Stuart Jordan, ed. NASA SP-450, Chapter 9.

Nicolson, I. 1982. *The Sun*. Mitchell Beazley Publishers, Mill House, London WIV 7 A.D.

Schatten, K. H. 1990. Climate Impact of Solar Variability, *EOS*, vol. 71, no. 33, p. 1103.

Shea, P. 1987. Solar-Planetary Relationships: Cosmic Rays, Solar and Interplanetary Physics, *Reviews of Geophysics*, vol. 25, no. 3, p. 641.

Sheely, N. R., Jr., Howard, R. A., Koomer, M. J., and Michels, D. J. 1983. *Association Between Coronal Mass Ejections and Soft X-Ray Events*, Naval Research Laboratory Publication 41-82-86.

Smith, R. E., and West, G. S. 1983. *Space and Planetary Environment Criteria Guidelines for Use in Space Vehicle Development 1982 Revision (Volume 1)*. NASA Technical Memorandum 82478, Chapter 1.

Spicer, D. S., and Brown, J. C. 1981. Solar Flare Theory, in *The Sun as a Star*, Stuart Jordan, ed. NASA SP-450, Chapter 18.

Suess, S. T. 1983. *Operational Uses for a Solar Soft X-Ray Imaging Telescope*, NOAA Technical Memorandum ERL-SEL-66.

Townsend, R. E. 1982. *Source Book of the Solar-Geophysical Environment*. Air Force Global Weather Central, Offutt AFB, NE, Chapters 7 and 8.

Wagner, W. J. 1987. Presentation to the USAF Scientific Advisory Board on Solar Flare Hazards to Man-in-Space, Space Environment Laboratory, Boulder, Col.

Webb, D. F., Krieger, A. S., and Rust, D. M. 1976. Coronal-ray Enhancements Associated with H-alpha Filament Disappearances, *Solar Physics*, 48, p. 159.

Zwaan, C. 1981. Solar Magnetic Structure and the Solar Activity Cycle Review of Observational Data, in *The Sun as a Star*, Stuart Jordan, ed. NASA SP-450, Chapter 6.

2.10   Problems

2-1.   The Sun generates energy at the rate 4×10^{33} erg/sec by the conversion of hydrogen to helium.

   a.   Show that if hydrogen could be completely converted to energy, it would require over 4 million metric tons of hydrogen to produce the observed solar energy generation rate.

   b.   Assume that the energy conversion reaction is simply

$$4H_1^1 \rightarrow He_2^4 + 2e^+ + 2\nu_e + 2\gamma + \text{energy}$$

   where e+ are positrons and the energy liberated per reaction is 25 MeV. Show that the Sun would need to convert 6.68×10^{11} kg of hydrogen into 6.6425×10^{11} kg of helium per second to produce the observed energy generation rate. (The mass defect for the energy conversion reaction is 0.0226 amu.)

2-2.   Differentiate between granulation, mesogranulation, and supergranulation on the Sun's surface. How deep into the Sun do these patterns penetrate? Are giant cells possible?

2-3.   Describe the principal solar magnetic surface features. Include such phenomena as pores, sunspots, fibrils, filaments, etc. Describe the life cycle of a sunspot group from birth to dissolution. Does the complexity of the sunspot group have any connection with flare activity? Explain.

2-4.   What is the solar neutral line and what does it have to do with solar flares? Explain. Does the neutral line have any association with coronal holes? Explain.

2-5.   In visible wavelengths would the solar atmosphere appear to become darker or brighter as we scan from the center of the solar disk to the outer edge (or limb)? Explain. How about at x-ray wavelengths? Explain. How about at microwave wavelengths? Explain.

2-6.   Graph the solar temperature and density profiles from the photosphere through to the corona. Are coronal electrons and protons gravitationally bound to the Sun? Explain. Is the corona a low or a high beta plasma? Explain the consequences of your answer. What are coronal loops and why are they important?

2-7.   During a solar eclipse we can see part of the solar atmosphere. Which part do we see and why? Should the features we see be symmetric around the Sun?

Does your answer depend on whether it is solar maximum or solar minimum? Explain.

2-8.   Show that the photospheric pressure is only 1% of the Earth's surface atmospheric pressure. Take the Sun's surface gravitational field to be 30 times as great as on Earth and the total mass of the solar atmosphere to be about 2.1×10^{19} kg compared to 5.29×10^{18} kg for the Earth's atmospheric mass.

2-9.   Assume that the average sunspot field is 0.3 tesla within a volume of $(10^4 \text{ km})^3$. Compute the percentage decrease in stored magnetic energy needed to produce a flare with energy of 10^{25} joules.

2-10. Why is a low beta plasma a requirement for most flare models? Do coronal mass ejections occur in low or high beta plasmas? Explain. How much magnetic energy needs to be converted into kinetic energy for plasma to reach escape velocity during a CME?

Chapter 3

Solar Wind

3.0 Introduction

The solar wind is basically a proton-electron gas that streams past the Earth with a mean velocity of 400–500 km/sec and a mean proton and electron density of about 5/cm³. This streaming solar plasma interacts with the Earth's intrinsic magnetic field and the resulting interaction has important repercussions on space systems operating in the near Earth environment. The only visible manifestation of this interaction is the aurora, which frequently lights up the sky at polar latitudes (both north and south). However, using suitable instruments one can observe any number of important space environmental and surface phenomena which are associated with the solar wind–geomagnetic field interaction.

The study of solar-terrestrial relations began in earnest during the first half of the twentieth century and considerable evidence was found for the presence of solar material in interplanetary space. "Solar corpuscular radiation" was widely invoked to explain polar aurorae, geomagnetic activity, and cosmic ray modulations. However, these explanations were in terms of streams of individual particles which did not interact with one another. The idea of a coherent stream of electrically neutral plasma leaving the Sun was developed in the 1930s by Chapman and Ferraro to explain the behavior of well-known geomagnetic disturbances. Later, Bartel postulated that this plasma stream was continuous but localized so the flow past the earth would be intermittent in agreement with the observed sporadic behavior of geomagnetic disturbances. In the 1950s these ideas were expanded by Biermann based on his study of the unusual behavior of comet tails in that they always point away from the Sun throughout their orbit about the Sun. Today, we know that the solar wind–comet tail interaction is far different from that proposed by Biermann. Nonetheless, these early studies played key roles in the development of our contemporary concepts of the solar wind. The real turning point in our understanding came through the in situ space observations of the 1960s. All reasonable doubt concerning the existence of an essentially continuous solar wind was removed in 1962 by nearly three months of continuous data taken by Mariner 2 on its voyage to Venus.

3.1 Theories—Historical Background

In the years following 1962, scientists have been very active in all phases of solar wind research. The solar wind exists because the Sun maintains a 2×10^6 K corona as its outermost atmosphere. At these temperatures the Sun is, in a sense, "boiling off" its outer atmosphere, and it is the continual expansion of the solar corona at supersonic speeds that we call the solar wind. This plasma wind, principally consisting of fully ionized hydrogen, has an average flow velocity of 400 km/sec (see Figure 3.1). Intermixed with this flowing solar wind plasma is the background coronal magnetic field which is trapped in or "frozen-in" the ever expanding solar wind (see Figure 3.2). As seen in Figure 3.3 these interplanetary magnetic field lines follow a spiral pattern which resembles the path traced out by water droplets flowing from a rotating water sprinkler. At Earth, the interplanetary magnetic field strength is about $5\gamma = 5 \times 10^{-5}$ gauss (about 10^{-4} times the Earth's surface magnetic field).

In 1958, E. N. Parker proposed that the solar coronal expansion was a natural result of the high temperature of the corona, and to illustrate this he worked on the first hydrodynamic model. The details of this model are not important for this discussion but his conclusion that the solar wind is a supersonic plasma was a radical departure from the popular "scientific truths" of the late 1950s. Remember at this time there were no in situ satellite measurements of the interplanetary medium. In contrast to Parker, J. W. Chamberlain contended that the solar corona expansion occurred at subsonic speeds and his model became popularly known as the "solar breeze." Ironically, both Parker and Chamberlain had part of the solution to the solar wind dynamics and our contemporary models borrow heavily from both of these earlier models.

3.2 Nozzle Analogy

The important characteristics of the solar wind can be demonstrated by a simple "gravitational nozzle" equation proposed by A. J. Dessler. In order to simplify the problem,

Figure 3.1 Coronagraph of Sun (6,400 Å) taken by the High Altitude Observatory at Boulder, Colorado. Notice the long coronal streamers which occur in areas which have an open magnetic field geometry allowing coronal plasma to escape *(courtesy Space Environmental Laboratory)*.

we'll assume that: 1) the solar wind can be treated as an ideal gas, 2) the solar wind flows radially away from the Sun, 3) the accelerations due to electromagnetic fields are negligible, and 4) the time scale over which the solar wind changes significantly is long compared to the time required for the solar wind generation (i.e., time stationary solutions). The equations we'll need are

$$\rho v r^2 = \text{constant} \qquad (3.1)$$

which is the conservation of mass for an outward expanding sphere.

$$\vec{F} = -\nabla p - \rho \vec{g} \qquad (3.2)$$

or in spherical coordinates,

$$\rho v \frac{dv}{dr} = \frac{-dp}{dr} - \frac{\rho GM_\odot}{r^2}$$

where $M_\odot$ is the mass of the Sun.

The speed of sound c_s can be related to pressure and density as follows:

$$c_s^2 = \frac{\dfrac{dp}{dr}}{\dfrac{d\rho}{dr}} = \frac{dp}{d\rho}. \qquad (3.3)$$

Combining equations 3.1, 3.2, and 3.3 yields

$$\frac{dv}{v}[v^2 - c_s^2] = \frac{dr}{r}\left[2c_s^2 - \frac{GM_\odot}{r}\right] \qquad (3.4)$$

or

$$\frac{dv}{dr} = \frac{v}{r}\frac{\left[2c_s^2 - \dfrac{GM_\odot}{r}\right]}{[v^2 - c_s^2]} \qquad (3.5)$$

where r increases radially outward. If

$$\frac{GM_\odot}{r} > 2c_s^2 \qquad (3.6)$$

and

$$v^2 < c_s^2 \qquad \text{(subsonic)} \qquad (3.7)$$

then

$$\frac{dv}{dr} > 0. \qquad (3.8)$$

This means that the solar wind accelerates outward, and the plasma flow (solar wind) can continue to accelerate and become supersonic ($v > c_s$). The transition from subsonic to supersonic flow occurs at the critical radius, r_c (where $v = c_s$). In order for a real, continuous solution to exist at r_c then

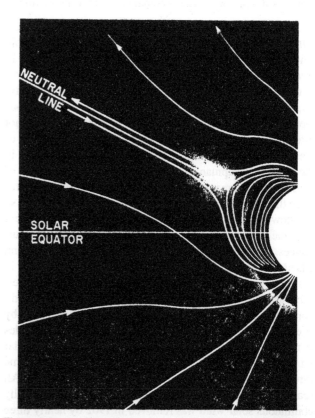

Figure 3.2 A hypothetical open magnetic field geometry is superimposed on a photograph of the white-light corona above the east limb of the Sun on June 5, 1973. Notice that as the dipole field lines are carried out into interplanetary space, the field direction reverses across the heliomagnetic equator—see section 3.6 *(after Hundhausen, 1978). A top view of the field geometry is shown in Figure 3.3.*

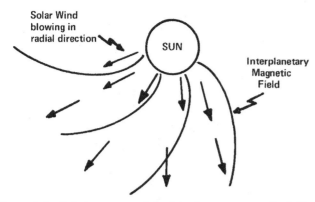

Figure 3.3 Spiral structure of the interplanetary magnetic field resulting from the fact that the ends of the field—being carried outward with the solar wind—remain attracted to the rotating Sun.

$$c_s = \sqrt{\frac{GM_\odot}{2r_c}} \qquad (3.9)$$

That is, at r_c the denominator of equation 3.5 goes to zero, and in order for equation 3.5 to remain mathematically valid then the numerator must also go to zero (i.e., the condition of equation 3.9). Therefore, the physical picture we have developed in the corona is that the solar wind remains subsonic until the condition expressed in equation 3.9 occurs and at such a location the solar wind can transition to supersonic speeds. Conversely, if the condition expressed in equation 3.9 never occurs, then the solar wind will always be subsonic.

3.3 Coronal Heating—Theory

Physically, the requirement of equation 3.6 is roughly equivalent to the situation of the gravitational potential energy being larger than the mean thermal energy. The gravitational potential energy is simply

$$P.E. = \frac{GM_\odot m}{r} \qquad (3.10)$$

where m is the mass of a solar wind particle, and the mean thermal energy of a gas is

$$TE = \tfrac{1}{2}m[\sqrt{2} \times 1.6c_s]^2 \approx 2mc_s^2 \qquad (3.11)$$

Therefore, the critical radius occurs in the region of the stellar atmosphere where the mean thermal energy is approximately equal to the gravitational potential energy. Returning to equation 3.5, one notices that if the solar wind is to continue to accelerate ($dv/dr > 0$) beyond the critical radius, then the mean thermal energy must exceed the gravitational energy (since $v > c_s$). The above model of a supersonic solar wind requires that a star have a cool lower atmosphere and a very hot outer atmosphere which has a heating mechanism that increases in strength with altitude.

How far should this outer atmospheric heating extend? In order to answer this question we need to look more at how the coronal gas responds to heating. We will also make the simplifying assumption that the coronal gas is ideal. The pressure of an ideal gas can be written as

$$P = k\rho^\gamma \qquad (3.12)$$

where

$$\gamma = \frac{c_p}{c_v} = \text{ratio of specific heat} \qquad (3.13)$$

and k is a constant. Equation 3.12 is referred to as the polytropic equation of state. When $\gamma = \tfrac{5}{3}$ the gas is said to be adiabatic (i.e., no heat gains or losses). Using the definition of equation 3.12, we can express equation 3.3 as

$$c_s^2 = \frac{\gamma p}{\rho}. \qquad (3.14)$$

The conditions of the cooler (lower) portions of our model atmosphere are summarized in equations 3.6, 3.7 and 3.8. If we now take the spatial derivative of equation 3.6 we find

$$0 > \frac{d}{dr} c_s^2 r \qquad (3.15)$$

or

$$0 > \frac{d}{dr}(r\rho^{\gamma-1}) \qquad (3.16)$$

where we have used equations 3.12 and 3.14. Solving equation 3.16 is straightforward and we find that in the lower atmosphere

$$\frac{\rho}{\rho_0} < \left(\frac{r_0}{r}\right)^{\frac{1}{\gamma-1}} \qquad (3.17)$$

where ρ_0 is the density at the base of the stellar atmosphere which occurs at radius r_0. If we now use the conservation of mass relation (equation 3.1) together with the equation of motion (equation 3.2) we find that

$$\frac{\rho}{\rho_0} > \left(\frac{r_0}{r}\right)^2 \qquad (3.18)$$

which is a second constraint equation (equation 3.17 being the first constraint). In order to satisfy both equations 3.17 and 3.18

$$\gamma < \tfrac{3}{2}. \qquad (3.19)$$

Therefore, the lower stellar atmosphere cannot be adiabatic ($\gamma = \tfrac{5}{3}$) and in order to maintain the outward expansion, heat *must be* provided from the cooler stellar surface. This may, at first, seem to be impossible. How can a relatively cool surface heat a contiguous atmosphere to temperatures far above the surface temperature? The answer is certainly not by thermal contact! But how about by the absorption of some type of wave energy?

One possibility is that the corona is heated by sound

waves, acoustic "noise" generated by turbulence at or just below the solar surface. The oscillatory motion of the Sun's surface would drive pressure waves into the overlying atmosphere in much the same way that hi-fi speakers drive sound waves into the air. The most popular version of this model called for heating the corona by shock waves generated as the sound traveled up into the progressively rarefied corona. In effect, the corona was heated by a succession of sonic booms. This model gained credibility in the early 1960s when observations showed that the surface of the Sun had strong vertical oscillations, just as the model required.

But the simple idea of coronal heating by acoustic waves has encountered criticism because experiments designed to observe the sound waves propagating up into the corona *have not* detected them. There is no question these waves exist at the surface and we also see them at somewhat greater heights where the temperature begins to climb. Yet the wave energy seems to diminish before it reaches the corona and is probably not the cause of the heating. Today, we feel that the heating mechanism may be due to the simple connection between magnetic fields and electric current.

The interior of the Sun acts as an electric dynamo which produces powerful magnetic fields. The surface fields of the Sun are always changing due to the continual emergence of new fields bubbling up from inside the Sun due to the turbulent overturning (convection cells) of the solar surface. This changing surface magnetic field induces an electric field which forces currents to flow in the highly conducting ionized solar atmosphere. Some of the current energy is lost due to resistance, and the lost energy heats the neighboring gas (ohmic heating). In its simplest form, the heating of the solar corona may be due to the same process which causes an ordinary electric light bulb to glow. However, when we examine the dynamics of the solar atmosphere in more detail we find that the above simple explanation is still not enough and more work needs to be done to uncover the details of the heating mechanism.

In summary, the outer layer of the solar atmosphere will accelerate outward given a suitable heating source which adds enough energy to overcome the Sun's gravitational potential energy. Gravity causes the atmospheric density to fall off quickly with a scale height approximately equal to $c_s^2 r^2/GM_\odot$. But remember, the atmosphere is being heated and because it is hot it will expand at a rate such that $\rho v r^2$ will be constant (conservation of mass). If the density, ρ, is falling off quickly then v must increase to maintain conservation of mass flow. This mechanism works because the atmosphere has a significant heat source at low altitudes. Without this heat source, the solar wind would never become supersonic. However there is a limit to how hot the atmosphere can be and still produce a supersonic stellar wind.

In order to understand how the solar wind depends on atmospheric temperature, we need to return to equation 3.5. For an ideal gas

$$p = nkT = \frac{\rho kT}{m} \qquad (3.20)$$

and

$$c_s^2 = \frac{dp}{d\rho} = \frac{kT}{m} \qquad (3.21)$$

where m is mass of gas particles.

If we substitute 3.21 into 3.5 we obtain

$$\frac{dv}{dr} = \frac{v}{r} \frac{\frac{2kT}{m} - \frac{GM_\odot}{r}}{v^2 - \frac{kT}{m}}. \qquad (3.22)$$

For very hot stars the numerator of equation 3.22 is always positive and the denominator is negative which means that as the atmosphere expands, the flow velocity slows down (dv/dr < 0) and never becomes supersonic. Conversely, for cool stars the numerator starts negative and since initially the outward flow rate, v, is less than the speed of sound, then the denominator is also negative. The result is that the flow accelerates outward (dv/dr > 0). After some time, v approaches the sonic velocity and at this location the amount of heating becomes critical. The flow will continue to accelerate only if the thermal energy equals and then exceeds the gravitational potential energy. Given these conditions, the flow velocity can continue to accelerate to high supersonic speeds (i.e., both the numerator and denominator are positive).

3.4 Interplanetary Medium

Our Sun (a cool star) appears to satisfy the necessary conditions for supersonic solar wind flow. Near the Earth's orbit, the solar wind's magnitude is in the range of 300–400 km/sec during quiet times, increasing to 600–700 km/sec in disturbed intervals. Plasma densities generally lie within the range of 3 to 40 particles/cm³ and kinetic temperatures from about 10^4 K to a few times 10^5 K. Intermixed with the streaming solar wind plasma is a weak magnetic field, called the interplanetary magnetic field or IMF. In the vicinity of the Earth, the IMF is about 1/10000 the strength of the Earth's surface magnetic field. On average the IMF is within the ecliptic plane (the plane defined by the Earth's orbit about the Sun) although, at times, the IMF can have a substantial component perpendicular to the ecliptic plane. As we will soon see, the direction of the perpendicular component determines whether the IMF will couple with the Earth's magnetic field which, in turn, allows solar wind particles easy access to the near Earth environment.

The hot coronal plasma has an extremely high electrical (as well as thermal) conductivity and as shown in section 1.2, the solar magnetic field becomes trapped ("frozen-in"). Therefore, the IMF is coupled to the continuous outward flow of coronal material which carries the solar magnetic field into interplanetary space. If the Sun did not rotate, the resulting magnetic configuration would be extremely simple: the radial coronal outflow of plasma would produce magnetic field lines extending radially outward from the Sun. Of course, the Sun does rotate with a period of approximately 25 days for a stationary observer (or approximately 27 days for an observer with the orbital motion of the Earth).

In a spherical coordinate system (r, θ, φ) rotating with the Sun (φ = 0 along the axis of solar rotation), the radial solar wind velocity components become

$$U_r = v = \text{expansion speed}$$

$$U_\theta = \omega r \sin \theta$$

$$U_\phi = 0$$

where $\omega = 2.7 \times 10^{-6}$ radians/sec is the angular velocity of solar rotation. The nonradial velocity component U_θ is entirely due to the Sun's rotation. Since the ends of the solar magnetic field lines remain embedded in the rotating Sun, the field lines carried radially outward by the solar wind are bent into something resembling Archimedean spirals (see Figure 3.4). The effect is similar to a rotating garden sprinkler where the jet of water follows a spiral path even though the individual drops are following radial trajectories. Just as the grooves of a phonograph record determine the path of the needle, so does the plasma determine the shape of the IMF. As seen from Earth (θ = 0), the spiral angle ψ (deviation from the radial direction) is given by

$$\tan \psi = \frac{r\omega}{v} \qquad (3.23)$$

where r is the radial distance from the Sun. Near the Earth, the spiral angle (sometimes called the "garden hose" angle) is about 45° under normal solar wind conditions (i.e., v ~ 400 km/s). However, due to eruptive disturbances on the Sun, variations in this spiral angle are common.

The radial and azimuthal IMF components near the ecliptic plane are given by

$$B_r = B_0 \left(\frac{a}{r}\right)^2 \qquad (3.24)$$

and

$$B_\phi = B_0 \left(\frac{a}{r}\right)^2 \frac{\omega}{v} (r - a) \qquad (3.25)$$

where B_0 is the radial field at the base of the corona (r = a), v is again the solar wind plasma speed, and ω is the angular velocity of the Sun. Therefore, a field of 1 gauss at the Sun's surface leads to about 3×10^{-5} gauss (3γ) IMF field strength near the Earth. The IMF can be directed either inward or outward with respect to the Sun, and perhaps the most remarkable result of the early satellite measurements was the tendency for large spatial regions (or temporal intervals) to have a well-defined predominant polarity (in or out from Sun) and for the polarity pattern to repeat on succeeding solar rotations. These regions are called "magnetic sectors" and, in a stationary frame of reference, the sectors would appear to rotate with the Sun. Typically, there are about four sectors and the boundaries between the magnetic sectors are shown schematically in Figure 3.5.

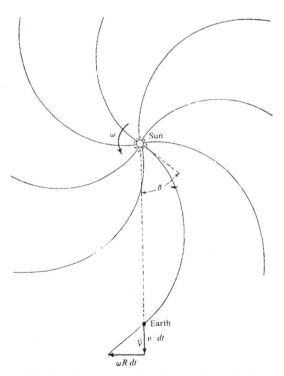

Figure 3.4 Graphic representation showing the angle ψ.

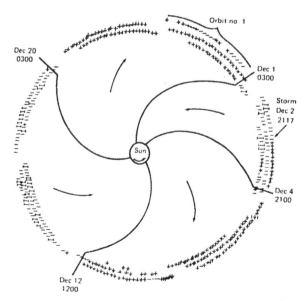

Figure 3.5 Solar magnetic sector structure inferred from the Interplanetary Monitoring Platform (IMP) spacecraft observations (Wilcox and Ness, 1965). Plus signs (away from the Sun) and minus signs (toward the Sun) are 3-hourly measurements of the interplanetary field direction, and the arrows in the four sectors indicate the predominant field direction.

The interplanetary sector pattern implies that the solar wind plasma within each magnetic sector emanated from a coronal region of similar magnetic polarity (i.e., retaining the identity of plasma and field lines because of the "frozen-in" conditions for the field). In interplanetary space, the kinetic energy density of the plasma is approximately 100 times the energy density of the magnetic field, so the IMF is clearly relegated to a passive role. Satellite data suggests that these magnetic sectors retain their basic identity for times longer than one solar rotation. A physical explanation for the interplanetary sector pattern is given in section 3.6.

3.5 Fast Streams and Coronal Holes

Since the solar wind is, in fact, part of the solar corona, it is to be expected that the sector structure of the solar wind will depend on the level of solar activity. Solar wind observations reveal large variations in solar wind speed and density on a time scale of several days. The basic physical feature of these so-called "fast" streams can be summarized as follows:

  a. The flow speed rises steeply from the pre-stream level (approximately 400 km/sec), reaching a maximum value (approximately 600 to 700 km/sec in about one day).

  b. The density rises to unusually high values (can exceed 50 cm^{-3}) near the leading edges of the streams and these high densities generally persist for about a day. These density peaks are generally followed by unusually low densities, persisting for several days.

  c. The proton temperature varies in a pattern similar to that of the flow speed.

The tendency for high-speed streams to recur on several successive solar rotations has not been clearly demonstrated, although it has been shown that these structures rotate with the Sun. The existence of a predominant magnetic polarity within high-speed plasma streams appears to show that the coronal magnetic structure plays an important part in the formation of these solar wind features.

The high-speed stream is also distorted into a spiral pattern by solar rotation and it will inevitably overtake and "collide" with any slower-moving ambient solar wind. The "frozen-in" conditions of the interplanetary plasma prevent the interpenetration of different streams, so that an interface boundary must form to separate fast streams from the ambient solar wind plasma. If the fast-stream plasma has a large relative velocity with respect to the ambient plasma, a shock front might be expected to form at the leading edge of the fast stream.

A model of the magnetic structure associated with the solar wind is shown in Figure 3.6. The region B_1 (high-speed stream) has coronal field lines which are open to interplanetary space. This configuration allows atmospheric gas to flow away from the Sun and the resulting low density region is known as a *coronal hole*, so named because these phenomena appear as vast dark areas on x-ray and ultraviolet photographs of the Sun. Typically, coronal holes occupy about 20% of the solar surface with maximum occurrence during solar minimum. There is gathering evidence that coronal holes are the major source of the recurrent high-speed streams in the solar wind. Surrounding the coronal hole field B_1 is the boundary field B_2 which separates the coronal hole region from the strong magnetic fields B_3 and B_4 of loop systems. These loop systems are normally remnants of active regions and because of their closed structure they can trap hot coronal plasma which appears "bright" on x-ray and ultraviolet photographs of the Sun (see Figure 3.7).

3.6 Interplanetary Current Sheet

The dipole nature of the main solar magnetic field adds latitudinal structure to the interplanetary medium. As the dipole field lines are carried out into interplanetary space, the field direction (i.e., away from or toward the Sun) reverses

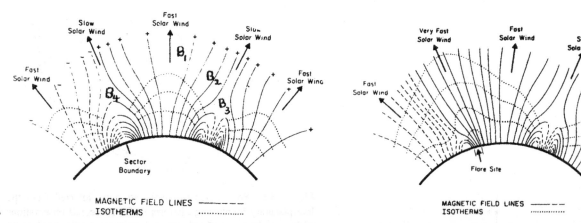

Figure 3.6   (A) A qualitative sketch of the theoretical coronal structure responsible for high-speed plasma streams. The symbols B_1, B_2, B_3, and B_4 are used to identify specific magnetic field features (see text). (B) A qualitative sketch of the temporary modification of the coronal structure (shown in A) due to flare activity. It has been assumed that the flare was sufficiently energetic to overcome magnetic forces and "open" a region of previously closed field lines *(after Hundhausen, 1978)*.

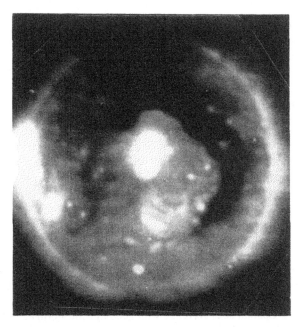

Figure 3.7 Soft x-ray photograph of corona taken during Skylab mission in May 1973. The dark areas are called coronal holes which have open magnetic flux tubes and are the source of the high speed solar wind streams. Evidence of coronal loops can be seen in the bright regions on the solar limb *(courtesy Space Environmental Support Center)*.

across the heliomagnetic equator. Physically, the oppositely directed field in each hemisphere has to be separated by a current sheet (sometimes called a neutral sheet—no magnetic field) of nearly one billion amperes. However, the current density is small considering that the current is spread over a disk of radius 100 AU. The current sheet prevents the oppositely directed magnetic fields from making contact and "explosively" destroying each other. The current sheet is inclined to the solar rotational equator by about 7° which causes it to have a wavy structure similar to Figure 3.8. Solar wind speeds are found to be a minimum in the immediate vicinity of the sheet, and increase with latitude above and below the current sheet. The corrugations (spacing between the waves in the current sheet) have an approximate wavelength of about 0.1 AU, and the average sheet thickness is about 3×10^4 km.

The inclination of the current sheet to the solar rotational equator ensures that the Earth will cross the current sheet at least twice during each solar rotation. Today, we believe that these crossings are the cause of the sector boundaries that we spoke of earlier. Such passages have a strong correlation with geomagnetic disturbances on Earth. However, eruptive disturbances (e.g., solar flares) on the Sun can modify the background sector structure, and during peak periods of solar activity, sector boundaries become undetectable.

During solar flares, the Sun may release MeV (million electron volts) electrons and protons as compared to average solar wind 0.25 eV electrons and 400 eV protons. (A room filled with 1 eV particles would have an effective kinetic temperature of over 11,000 K.) The time required for these

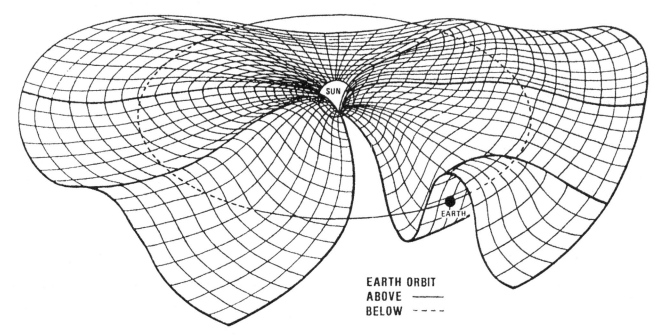

EARTH ORBIT
ABOVE ———
BELOW ‑ ‑ ‑ ‑

Figure 3.8 Wavy structure of the heliomagnetic current sheet due to current sheet's inclination to the Sun's rotational equator *(after National Research Council, 1981).*

high energy particles to reach the Earth depends on flare location, particle energy (speeds), and the shape of the field line emanating from or around the flare region. The flare ejecta and its accompanying "frozen in" solar magnetic field cannot cross the current sheet (although it may alter its position), and therefore it is very important to know where the Earth and flare source regions are with respect to the current sheet. For this reason, attempts are made to trace the sector boundary back to the solar surface. During solar cycle 21, the field was directed toward the Sun when the Earth was below the current sheet, and was directed away from the Sun above the current sheet.

3.7   Solar Wind Power Generation

As we will see in Chapter 5, the solar wind drives the dynamics of the Earth's magnetosphere. One way to look at the solar wind-magnetosphere interaction is as an electrical circuit problem. Since the solar wind consists of charged particles in motion, one might ask what is the electrical power potential of the solar wind to drive the complex current driven energy dissipation mechanisms occurring in the magnetosphere. As one might imagine, a detailed analysis of such interactions is extremely complex and poorly understood. But in simple terms we can at least determine what factors would go into determining the solar wind's electrical power potential. From elementary physics the power available in a simple circuit is just

$$P = VI \qquad (3.26)$$

where P is the power, V is the electric potential, I is the available current. In Chapter 1 (see equation 1.16) it was stated that the condition for the solar wind to exhibit frozen-in-flux is

$$\vec{E} + \vec{v} \times \vec{B} = 0.$$

This means that the magnitude of the solar wind's electric field is proportional to the velocity, v, times the magnetic field strength, B; that is, $E \propto vB$. Once again, from elementary circuit theory the electrical potential is simply

$$V = \int_x^y \vec{E} \cdot d\vec{l} \qquad (3.27)$$

where the voltage is measured between points X and Y along the integral path. For a constant electric field, the electric potential is simply

$$V \propto Ed \propto vB \qquad (3.28)$$

and the power obeys the following proportionality:

$$P \propto vBI. \qquad (3.29)$$

In section 1.6 we showed that for a gyrating particle the equivalent current is given by equation 1.30 as

$$I_e = \frac{q\omega_c}{2\pi} = \frac{q^2 B}{2\pi m}$$

and the total equivalent current is the sum of the particles crossing between points X and Y. Therefore, the power can be expressed as

$$P \propto vB^2 L_0^2 \qquad (3.30)$$

where L_0 is a characteristic length, taken to be the radius of the cross-sectional area between points X and Y.

More detailed treatments of magnetospheric energy dissipation have revealed that the power input rate is of the form (proportional to)

$$\epsilon \propto vB^2 \sin^4(\theta/2) L_0^2 \qquad (3.31)$$

where θ describes the orientation of the Z component of the interplanetary magnetic field, B. When B_z is northward (up) θ is equal to $0°$; when B_z is southward, θ is $180°$. The so called epsilon parameter, ϵ, is the key parameter in the "directly" driven model of the magnetosphere (see section 5.8). Epsilon does correlate with geomagnetic storm activity but the correlation is far from perfect (see problem 5-10) because all the mechanisms involved in coupling the available input energy of the solar wind into the magnetosphere are still poorly understood. Furthermore, we are still discovering the many energy dissipation processes which exist in the magnetosphere.

3.8   Libration Point Measurements

Over the past 20 years our knowledge of the solar wind has grown tremendously by in-situ measurements from the Pioneer, Voyager, Viking, Mariner, and Explorer series of interplanetary spacecraft. These probes have provided a lengthy history of the solar wind measurements as they flew to Mercury, Venus, Mars, Jupiter, Saturn and beyond. Their measurements have extended over two solar cycles and covered a huge expanse of space. However, it was not until International Sun Earth Explorer 3 (ISEE 3) reached a libration point parking orbit that we were able to gain detailed information on the solar wind–magnetospheric interaction. This was possible because the satellite's orbital position allowed scientists to sample the solar wind just before it impacted the Earth's magnetosphere. ISEE 3 remained at the libration point for about 4 years from 1978 to 1982, after which it moved off into an intercept orbit to rendezvous with Halley's comet. A libration point orbit of this type is unique in that the satellite maintains its same relative position between the Sun and the Earth.

From classical mechanics, the restricted three body problem shows that the orbit of a very small body m, moving in the gravitational field of objects of mass M_1 and M_2, has five equilibrium locations where all forces acting on m balance to zero. Therefore, a satellite could be "parked" at one of these equilibrium points, thereby keeping the same relative position between M_1 and M_2 given that one of the objects (M_1 or M_2) has most of the total mass of the system. However, even small perturbations can displace a satellite from one of these points, and thus the satellite would rock back and forth about the equilibrium position, or in the terminology of orbital mechanics, the satellite would "librate." In some textbooks the equilibrium points are referred to as Lagrangian points. One of these orbital equilibrium points, sometimes called L1, lies along the line connecting M_1 and M_2. In the case of the Sun–Earth system, the L1 point parking orbit is approximately 1.5 million kilometers sunward of Earth, well outside the Earth's bow shock. In fact, under normal solar wind conditions, it takes solar wind particles about 1 hour to cover the distance from the libration point to the Earth.

The gravitational force of a fourth body, the Moon, complicates the orbital dynamics of an L1 satellite. As a result, a libration satellite actually orbits around the L1 point, as the L1 point itself orbits the Sun. This motion within an orbit is sometimes referred to as a halo orbit, and its radius depends on the strength of the perturbing forces. In the case of ISEE-3, the halo orbit is elliptical with a semimajor axis length approaching 6.4×10^5 km and a period of about 178 days. The halo orbit's major axis is orthogonal to the Sun–Earth line. Therefore, a libration point satellite is in an outstanding location to gather both valuable research data and extremely useful operational forecasting information.

From an operational forecasting perspective, an L1 satellite is an important part of an overall space forecasting system. The forecast starts with Earth-based solar telescopes which monitor solar activity in a variety of wavelengths including optical, radio, and soft x-ray imagery. These observations provide the first warning of potential problems and allows a forecaster to make a "first guess" forecast based on flare location and intensity. However, in order to refine the forecast, more detailed information about the solar ejecta speed and direction is needed. The ideal instrument for such measurements is one which can focus on the coronal plasma, for example, a solar coronagraph.

Typically, a coronagraph isolates the corona by using an occulting disk to block out the electromagnetic radiation from the photosphere and chromosphere. By using a coronagraph aboard a satellite, scientists and forecasters could watch CMEs and flare ejecta move away from the Sun. The direction of the solar plasma is determined from the pattern shape of the plasma as it moves away from the Sun. Symmetric patterns would indicate the plasma is moving along the Sun–Earth line whereas asymmetric patterns indicate the plasma is moving off at some angle to the Sun–Earth line. Solar coronagraphs have been flown on two satellites: P78-1/Solwind and the Solar Maximum Mission (SMM). The Solwind satellite could see out to 10 solar radii and SMM had a range of 6 solar radii. As a matter of comparison, ground-based observations can see out to 2–3 solar radii. Both satellites were designed for research purposes and, therefore, did not provide real-time operational data.

Finally, the L1 satellite would be the last link in the forecasting system. If the solar plasma ejecta had an intersecting trajectory with respect to Earth, the L1 satellite would sample the solar wind plasma density, speed, and magnetic field orientation shortly before impacting the magnetosphere. Thus these measurements would give forecasters precisely the information they would need to predict the onset time of the geomagnetic storm, and would also help gauge the intensity and duration of the storm. While at L1, ISEE 3 measurements were made available to the forecast community and the data did prove to be an extremely effective short-term forecast tool.

3.9 Supersonic Solar Wind: A Contradiction

Before we end this chapter, we need to clarify the expression, "the solar wind is supersonic." Normally, when we talk about a high density gas, the mean free path is a critically important parameter. The distance between collisions limits how fast information can travel through gas by kinematic processes and this limiting speed is called the sonic velocity. Any process exceeding this limit is no longer continuous and at the sonic/supersonic boundary there is a jump in gas characteristics—such as density, temperature, etc.—across the boundary (see section 5.1). We successfully applied these ideas for the solar corona in section 3.2. However, the mean free path for the solar wind *near the Earth* is nearly 1 AU, so how can the fluid methodology apply? For such a large mean free path, is the concept of sonic velocity even meaningful for the solar wind?

In the early 1960s, aerodynamic models were applied directly to the magnetosphere. People reasoned that since the solar wind is a high beta plasma, the fluid flow should govern the magnetospheric interaction. Even though the solar wind is basically collisionless, the classical aerodynamic models reasonably reproduced the shape of the Earth's bow shock (see Figure 5.1) for an upstream mach number of 8 or more. Again, mach number is the ratio of fluid flow velocity to the speed of sound in the fluid. Of course, the speed of sound assumes that the fluid is thermalized ($\gamma = \frac{5}{3}$ for an ideal gas). Near the Earth, the solar wind plasma is anything but thermalized. The solar wind is highly ordered in direction and the principal fluid pressure is the dynamic pressure (ρv^2) not the gas or thermal pressure. However, there must be some physical reason why classical fluid models describe so well the solar wind behavior in the vicinity of the Earth.

The mean free path is just the interaction distance for individual fluid particles; it is a parameter that is based on the fluid particle's range—the distance the particle can move before it changes direction. Although a low density, magnetized plasma has very few (if any) collisions, the individual, charged particles are limited in range, not by collisions, but, instead, by the gyroradius. In simple terms, the gyroradius acts as an effective interaction length. Remember, in a plasma a charged particle can make its presence felt without colliding with another charged particle. For example, momentum between charged particles can be exchanged through electromagnetic fields. The acceleration experienced by a gyrating charged particle produces electric and magnetic fields which can influence other charged particles. Think of a plasma as consisting of particles on a string (the magnetic field line). If some mechanism makes the string shake, then all the beads are affected even though they do not actually collide with each other.

The gyroradius is given by equation 1.39 and is repeated here for convenience.

$$r_c = \frac{mv_\perp}{qB} \qquad (3.32)$$

where $v_\perp$ is the velocity component perpendicular to the orientation of B and q is the charge of a proton. The effective collision frequency is simply the gyrofrequency

$$\nu_c = \frac{\omega}{2\pi} = \frac{qB}{2\pi m}. \qquad (3.33)$$

In this new pseudo gas, the speed of sound is on the order of

$$c_s \sim r_c \nu_c \sim \frac{v_\perp}{2\pi} \qquad (3.34)$$

and the mach number becomes

$$M = \frac{v_\perp}{c_s} \sim 2\pi. \qquad (3.35)$$

Therefore, the gyromotion makes the plasma act as a pseudo-supersonic fluid.

But, is there any real physical mechanism that can carry information between particles in the plasma? Of course such a process cannot depend on collisions and must therefore rely on electromagnetic fields. As we mentioned earlier, the magnetic field interweaves the individual charged particles making up the plasma. An overused, but sometimes useful analogy is to treat the magnetic field lines as a stretched string or rubber band. Let me warn you that this analogy has limited use because magnetic fields are not strings nor rubber bands, and the only correct analysis requires a determination and time development of the pertinent current systems. With this warning in mind, we will use the stretched string analogy in an example where it *does* work. We all know that waves can travel on a string, so in a sense, we can send information along the string because the wave movement replaces collisions as the carrier of information to the particles on the string. From classical physics the velocity of the wave depends on the tension of the string and the inverse mass density. That is,

$$v_T = \sqrt{\frac{T}{\rho}} \qquad (3.36)$$

where T is the tension and ρ is the mass density per unit length. In Chapter 1 we saw that the magnetic field tension is given by B^2/μ_o so that the velocity of waves traveling along a uniform B field should be of the form

$$v_A = \sqrt{\frac{B^2}{\mu_o \rho}} \qquad (3.37)$$

where these traveling "magnetic waves" are usually called Alfven waves. Therefore, the Alfven velocity, v_A, is the limit at which information can be carried in our collisionless plasma. Compared to the solar wind flow velocity (v), we have

$$\frac{v}{v_A} \sim 10. \qquad (3.38)$$

The above discussion shows that in the vicinity of the Earth the solar wind is truly superalfvenic and can, in a sense, be treated as a supersonic fluid. But remember, that the terminology and algorithms for supersonic flow really belong in the realm of a collisional, thermalized gas. However, the concept of supersonic flow can be applied to the solar wind near the Earth if the reader is cautious and aware of its limitations.

3.10 Summary

A continuous flow of protons and electrons, called the solar wind, streams away from the Sun with a mean velocity of 400–500 km/sec, and a mean proton and electron density of 5 particles per cm³. The origin of the solar wind is within the extremely hot, outermost atmospheric region of the Sun, the corona. The high electrical conductivity of the solar wind locks ("frozen-in conditions") the coronal magnetic field with the outflowing solar wind gas. The resulting interplanetary magnetic field has a pinwheel-like appearance and has a field strength of about 5 gammas near the Earth. Furthermore, an interplanetary current sheet must exist in order to maintain the stretched interplanetary magnetic field geometry.

3.11 References

Akasofu, S.-I. 1982. Interaction Between a Magnetized Plasma Flow and a Strongly Magnetized Celestial Body With an Ionized Atmosphere: Energetics of the Magnetosphere. *Annual Review of Astronomy and Astrophysics*, vol. 20, p. 117.

Bate, R. R., Mueller, D. D., and White, J. E. 1971. *Fundamentals of Astrodynamics*. Dover Publications, New York, Chapter 7.

Donnelly, R. F. ed. 1979. *Solar-Terrestrial Predictions Proceedings*. U.S. Department of Commerce, Boulder, Col.

Falthammer, C. G. 1973. The Solar Wind, in *Cosmical Geophysics*, Egeland et al., eds. Universitetsforlaget, Oslo Sweden, Chapter 7.

Glasstone, S. 1965. *Sourcebook on the Space Sciences*. D. Van Nostrand Company, Inc., Princeton, N.J., Chapter 9.

Hargreaves, J. K. 1979. *The Upper Atmosphere and Solar-Terrestrial Relations*. Van Nostrand Reinhold Co., New York.

Hill, Thomas W., and Wolf, Richard A. 1977. Solar-Wind Interactions in *The Upper Atmosphere and Magnetosphere*. National Sciences, Washington, D.C., p. 25.

Hollweg, J. V. 1981. The Energy Balance of the Solar Wind, in *The Sun as a Star*. NASA Monograph, NASA SP-450.

Hundhausen, A. J. 1972. *Coronal Expansion and Solar Wind*. Springer-Verlag, New York.

Kahler, S. 1987. Coronal Mass Ejections. *Reviews of Geophysics*, vol. 25, no. 3, p. 663.

Kopp, R. A. 1981. Heating and Acceleration of the Solar Wind, in *The Sun as a Star*. NASA Monograph, NASA SP-450.

Kuperus, M., Ionson, J. A., and Spicer, D. S. 1981. On the Theory of Coronal Heating, *Annual Review of Astronomy and Astrophysics*, vol. 19, p. 7.

National Research Council. 1981. *Solar-Terrestrial Research for the 1980s*. National Academy Press.

Nishida, A. 1978. *Geomagnetic Diagnosis of the Magnetosphere*. Springer-Verlag, New York.

Piddington, J. H. 1981. *Cosmic Electrodynamics* (2nd Ed). Robert E. Krieger Pub. Co., Malabar, Fla.

Schulz, M. 1976. Plasma Boundaries in Space. Air Force Systems Command Report SAMSO-TR-76-220.

Schwenn, R., and Rosenbauer, H. 1984. 10 Years of Solar Wind Experiments on HELIOS 1 and HELIOS 2, in *10 Years of HELIOS*. H. Porsche, ed., Max-Planck Institute, Germany, p. 66.

Space Science Board. 1978. *Space Plasma Physics; The Study of Solar-System Plasmas*. National Academy of Sciences, Washington, D.C.

Smith, R. E. and West, G. S. 1983. *Space and Planetary Environment Criteria. Guidelines for Use in Space Vehicle Development, 1982 Revision (Volumes 1 and 2)*. NASA Technical Memorandum 82 478.

Symon, K. R. 1960. *Mechanics, 2nd Edition*. Addison-Wesley Publishing Co., Reading, Mass., Chapter 7.

Thomas, Barry T., and Smith, E. J. 1981. The Structure and Dynamics of the Heliospheric Current Sheet. *J. Geo. Res.*, preprint.

Vasyluinas, V. M. 1974. Magnetospheric Cleft Symposium. *Trans. Amer Geo Union*, 55, p. 60.

Wentzel, D. G. 1981. Coronal Heating, in *The Sun as a Star*. NASA Monograph, NASA SP-450.

Wilcox, J. M. and Ness, N. F. 1965. Quasi-Stationary Corotating Structure in the Interplanetary Medium. *J. Geophys. Res.*, 70, 5793.

Wolf, R. 1979. *Plasma Physics*, unpublished book/class notes. Rice University, Houston, Tex.

3.12 Problems

3-1.   Compare the physical characteristics of the corona with the solar wind. How does the solar wind speed and density vary as a function of distance from the Sun? How does the magnitude and orientation of the interplanetary magnetic field vary as a function of distance from the Sun?

3-2.   Suppose, using the space telescope we detect that a nearby hot star has a stellar wind flowing at 600 km/sec. Could we say that this stellar wind is supersonic? Explain.

3-3.   If the solar wind speed doubles, how will the spiral (or "garden-hose angle") change? Does this situation ever occur? Explain.

3-4.   Compare the solar wind particle density at the time of solar minimum with the particle density at solar maximum. Does the solar wind speed change during the solar cycle? Explain. Does the interplanetary magnetic field strength vary with the solar cycle? Explain.

3-5.   What are coronal holes and why are they of potential importance for the solar wind? Would a disappearing solar filament be of any importance for the solar wind? Explain. Can we detect changes in the solar wind speed from ground based detectors? Are space-based detectors necessary, and, if so, why? Can we detect these changes close to the Sun or do we have to wait until the solar particles arrive at Earth?

3-6.   What does it mean to say that the "solar wind is supersonic"? Supersonic compared to what? Compute the sonic velocity of the plasma making up the corona. Does the sonic velocity change with the solar cycle? Explain.

3-7.   Waves can travel along magnetic field lines much like waves travel along a stretched string. The speed of these "magnetic" waves (called Alfven waves) is given by equation 3.37 as

$$v_A = \sqrt{\frac{B^2}{\rho\mu_o}}$$

where B is the magnetic field strength, ρ is the plasma mass density, and μ_o is the magnetic permeability of free space. Is the solar wind superalfvenic (i.e., faster that v_A)? If so, why is this important? Is this in any way analogous to being supersonic? Compute the Alfven speed of the plasma making up the corona. Is the solar wind superalfvenic at the sonic/supersonic transition point? Explain. Where is the superalfvenic transition point?

3-8.   Describe why the Sun needs a heliomagnetic current sheet. How long does it take the Earth to cross this current sheet? How far away from the Sun should we expect the current sheet to extend? Can you think of an experimental method to verify your answer?

3-9.   Describe why the wavy pattern of the heliomagnetic current sheet can help us explain the observed solar sector structure. How many waves are in a typical current sheet pattern? Does this number vary with the solar cycle? Do solar flares or CMEs affect the current sheet itself or the current sheet pattern?

Chapter 4

Geomagnetism

4.0 Introduction

Over 2000 years ago, it was known that stones, found by Greeks in the region Magnesia, would attract pieces of iron. In the Middle Ages these stones, called lodestones (today we call this mineral magnetite), were placed on a block of cork or wood; floating freely on still water, the piece of lodestone would align itself in approximately a north-south direction and thus become a simple magnetic compass. In the sixteenth century, William Gilbert, Queen Elizabeth's physician, suggested that a compass always points north and south because the Earth has intrinsic magnetic properties. By the mid-nineteenth century, scientists (Oersted and Ampere to name a few) had proposed that electric currents are the source of all magnetism, even geomagnetism.

As measurements became more exact, the magnetism at any one point on the Earth's surface was found to vary, both slowly over the years (secular variation) and more rapidly, from month to month, day to day, even hour to hour (called transient variations). Some of these rapid variations are known to be periodic and regular, and associated with the positions of the Sun and Moon. The solar daily variations occur because, at high levels in the atmosphere, the Sun's shortwave radiation ionizes oxygen and other molecules, and forms a conducting layer, the ionosphere, mainly on the sunlit side of the Earth (see Chapter 7). Over the sunlit hemisphere there are also solar thermal and tidal motions, as well as lunar tidal motions, in the atmosphere that generate currents in the electrically conducting ionosphere. These currents decrease at night because the intensity of the conducting ionosphere diminishes steadily at night. This process is fairly regular, though it does vary from day to day and season to season because of the changing geometrical relationships between the Earth, Sun, and Moon. However, there remain frequent non-periodic short-term variations in the main geomagnetic field due to solar induced disturbances in the distant regions of the Earth's field. In 1852, Sir Edward Subine related increased magnetic activity with unusually intense solar flares. Today we know the large amounts of electromagnetic energy emitted during flares add additional ionization to the Earth's upper atmosphere. The resulting enhanced electrical conductivity changes the upper atmospheric current patterns which, in turn, produces irregularities on surface measurements of the Earth's main field. After a period of up to a few days, the charged particles expelled by the flare may also arrive at Earth, and their interaction with outermost parts of the geomagnetic field can produce additional currents which also register as irregularities on surface field measurements. Today, there is a worldwide network of stations making routine geomagnetic measurements.

4.1 Dipole Field

To a first approximation, the Earth's magnetic field is that of a sphere uniformly magnetized in the direction of the centered dipole axis. This axis cuts the surface of Earth at two points, A and B, known as the austral (south) and boreal (north) dipole poles. The "best fit" (smallest cumulative error) between the Earth-centered dipole and the actual magnetic field is obtained by taking A at 78.5°S–111°E (near Vostok Station, Antarctica) and B at 78.5°N–69°W (near Thule, Greenland). These points are about 800 miles from the geographic poles and are called the "geomagnetic poles." It will be seen that the axis of the dipole does not coincide with the axis of rotation, the angle of displacement being 11.3°. The plane through the center of Earth perpendicular to the dipole axis is called the dipole equatorial plane, and the circle in which it cuts the sphere is called the dipole equator. Dipole latitude is reckoned relative to this equator and the semicircles joining B and A are called the dipole meridians. The relationships between the dipole coordinates (latitude Φ and longitude Λ) and the corresponding geographic coordinates (ϕ, λ) at a point, P, are given by

$$\sin \Phi = \sin \phi \, \sin \phi_o + \cos \phi \, \cos \phi_o \, \cos (\lambda - \lambda_o) \quad (4.1)$$

$$\sin \Lambda = \frac{\cos \phi \, \sin (\lambda - \lambda_o)}{\cos \Phi} \quad (4.2)$$

where ϕ_o and λ_o are the geographical latitude and longitude of the north dipole pole ($\phi_o = 78.5°N$, $\lambda_o = 291.0°E$).

The magnetic potential, V, of a dipole with magnetic moment M is

$$V = \frac{\vec{M} \cdot \vec{r}}{r^3} = -\frac{M \sin \Phi}{r^2} \quad (4.3)$$

where Φ denotes the dipole latitude, and r is distance from the center of the Earth. M for the Earth is $8.05 \pm 0.02 \times 10^{25}$ gauss-cm^3.

The radial (vertical) component, Z, of the field is given by

$$Z = \frac{\partial V}{\partial r} = \frac{2M \sin \Phi}{r^3}. \qquad (4.4)$$

The horizontal (tangential) component, H, is given by

$$H = \frac{1}{r} \frac{\partial V}{\partial \Phi} = -\frac{M \cos \Phi}{r^3} \qquad (4.5)$$

and the magnetic dip or inclination, I, is given by

$$\tan I = \frac{Z}{H} = -2 \tan \Phi. \qquad (4.6)$$

Also, the magnetic intensity, B, is derived from

$$B^2 = H^2 + Z^2. \qquad (4.7)$$

At the pole, B = Z, and at the equator, B = H. Let H_o denote the equatorial value of H at the surface of the sphere (r = a), then we have

$$H = H_o \left(\frac{a}{r}\right)^3 \cos \Phi \qquad (4.8)$$

$$H = 2H_o \left(\frac{a}{r}\right)^3 \sin \Phi \qquad (4.9)$$

$$H = H_o \left(\frac{a}{r}\right)^3 \{1 + 3 \sin^2 \Phi\}^{1/2} \qquad (4.10)$$

It will be seen from equations 4.8, 4.9, and 4.10 that the magnetic intensity at the poles is twice that at the same radial distance (height) above the equator. Notice that the strength of the magnetic field decreases as the cube of the distance from Earth's center. Thus at 1 R_E above the surface, the field is only 0.125 of that at the surface.

The lines of force of the dipole field are given by dr/Z = rdΦ/H, and their equation becomes

$$r = ka \cos^2 \Phi. \qquad (4.11)$$

The product, ka, is the distance at which the lines of force cross the equatorial plane (a = surface radius). Therefore, the points where a line of force meets the sphere are given by $\cos \Phi_o = k^{-1/2}$.

On Earth, the direction of H is specified by the angle, D, between H and the geographic north; D is called the magnetic (or compass) declination and is reckoned positive if eastward. The northward and eastward components of H are denoted by X and Y, respectively, and

$$\tan D = Y/X \qquad (4.12)$$

$$X = H \cos D \qquad (4.13)$$

$$Y = H \sin D. \qquad (4.14)$$

The seven quantities, B, Z, H, I, D, X, and Y (see Figure 4.1) are called magnetic elements and any set of three independent elements serves to specify B; i.e.,

B, I, D; H, I, D; H, Z, D; X, Y, Z.

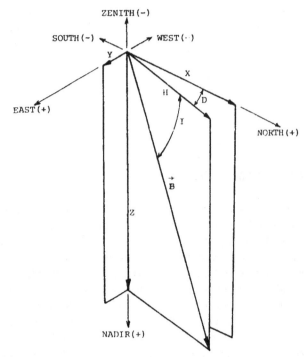

Figure 4.1 Elements of the Earth's magnetic field.

The elements of B, H, Z, and I are called intrinsic because their reference direction is the magnetic field itself. The other three elements are called relative because their reference direction is the geographical, north-south axis which may or may not be related to the magnetic field direction.

In geomagnetism the field intensity is usually measured in units of gauss. The smaller unit, γ, is used in conjunction with the variations in the geomagnetic field and

$$1 \gamma = 10^{-5} \text{ gauss.}$$

From equation 4.10, the surface geomagnetic field is about 0.6 gauss at the polar regions and about 0.3 gauss in the equitorial regions.

The distribution of Earth's magnetic potential, V (from which the vector field can be derived) can be expressed by a mathematical process called spherical harmonic analysis. The potential, V, can be expressed as a sum of orthogonal functions (called seminormalized associated Legendre functions of order m and degree n) of the type:

$$P_n^m (\cos \theta) (\alpha_n^m \cos m\lambda + b_n^m \sin m\lambda) \qquad (4.15)$$

where θ is the north co-latitude (i.e., the angular distance from the north pole) and λ is the east longitude. The functions P_n^m have been tabulated. By use of this technique, it has been shown that most of the field originates from within Earth. A small part (less than 0.1%) has its origin outside Earth due to ionospheric currents, and these ionospheric currents vary in intensity due to changes in solar activity (see

Chapters 5 and 8). Figure 4.2 shows the global distribution of magnetic field strength computed by the above method.

4.2 The Main Field of the Earth

The source of the Earth's internal magnetism is not permanently magnetized minerals in the Earth because our planet is internally much too hot for any magnetic material to retain its magnetism. Also, permanently magnetized minerals cannot move about rapidly enough to account for the known long-term changes in the strength, direction, and surface variations of the Earth's magnetic field. Seismic measurements indicate that at least part of the Earth's core is a metallic-like fluid which surrounds a denser (probably solid) inner core. It is almost universally believed that the fluid motions of the liquid portions of the core generate currents, which in turn induce the magnetic field. However, there is little agreement on what energy source drives the fluid motion and on how that motion gives rise to magnetic fields.

Rocks in the Earth's crust provide information about the history of the geomagnetic field. Studies indicate that the Earth has had a significant field for at least 2.7 billion years (the Earth is at least 4.5 billion years old), and that the field reverses direction every 10^5–10^6 years. The general north-south alignment of the field suggests that the field is mainly dipolar but it certainly is not a perfect dipole. Historical records show over the past four centuries that the geomagnetic field is decreasing in strength at a rate which will eliminate the field in about 3000 years, and the records also show a slow westerly drift of the magnetic poles. Any hypothesis about the source of the geomagnetic field must be able to explain the above phenomena.

For the past two decades, the most popular geomagnetic field theory was proposed by Elasser and Bullard in the 1950s. This so-called self-sustaining dynamo model converts the mechanical motions of the core materials into electrical currents. The model assumes that the Earth was formed with a small initial magnetic field (maybe the residual field which permeates the entire galaxy) which then became the seed field to start the dynamo process. Once started, the dynamo will generate its own field without any external supply of magnetism. However, the dynamo does require a constant source of mechanical energy to keep the metallic core material moving.

Rotation clearly must play a crucial role in the formation of geomagnetism because other rapidly rotating planets, the Sun, and other rotating stars all seem to have magnetic fields that are correlated to their axis of rotation. A possible explanation uses the Coriolis force that acts on the rotating core material. In the atmosphere and the oceans the Coriolis force is responsible for the large-scale cyclonic atmospheric

Altitude 0 km, Lines of Constant B, gauss

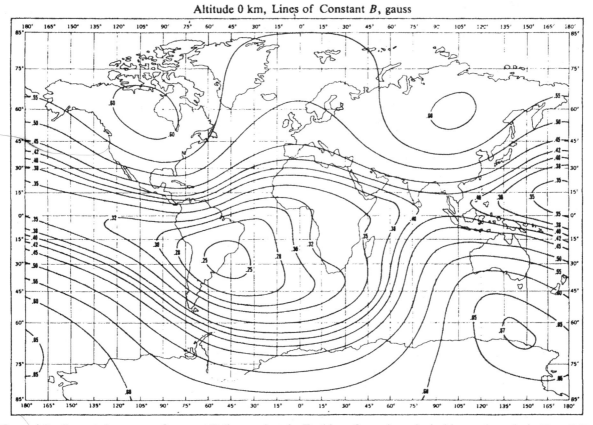

Figure 4.2 Computed contours of constant B (in gauss) at the Earth's surface using spherical harmonic analysis *(Hess, 1968).*

motions and the sea currents. However, the Coriolis force can only change the direction of currents; it cannot change the speed of the flow because it only acts perpendicular to the direction of the flow. Therefore, once motion begins the Coriolis force helps shape the pattern of the flow. Gravity probably drives the motion because it pulls the denser core material downward, forcing the lighter material upward. However, the simple exchange of material will not persist forever. Some source of energy must be provided to keep this "buoyancy" circulation going.

One possible mechanism is a continuous source of heat. A possible, although not well accepted, heat source is from large amounts of radioactive elements within the core material. Another possibility is that the core is cooling and as the core changes from liquid to solid it releases significant amounts of latent heat. The thermal capacity of the core is so great that a steady temperature drop of 100 K over the past 3 billion years would provide enough heat. Due the poor (estimated) thermal efficiency of the core, it is generally believed that the above heat driven dynamo could at best generate only a small magnetic field.

Another possible candidate is the gravitationally powered dynamo. Seismic measurements indicate that there may be a 20% difference in the density between the solid material of the inner core and the liquid of the outer core. As the liquid cools and "freezes" out metals (such as iron), they will probably migrate into the solid inner core as the lighter materials stay behind and float to the outside of the liquid core. The gravitational energy that is lost through this redistribution can be converted efficiently into heat which in turn can drive the dynamo. Because of where and how the heating is produced, most of the gravitational energy goes into generating a magnetic field. Estimates are that a gravitational energy source could generate magnetic fields of hundreds of gauss without too much heat loss into the mantle (the principal means of efficiency reduction).

Even though the configuration of these dynamo currents is such that the field at the surface of Earth resembles that of a dipole, there are appreciable departures from the dipole field which are called "anomalies." The large or regional anomalies affect areas of thousands of square kilometers; they are attributed to irregularities or eddies in the internal current system, probably near the boundary between the core and the mantle, so that their effects are most pronounced at those portions of the surface immediately above them. These large anomalies are moving westward very slowly, which indicates that the rates of rotation of the core and the crust of Earth are slightly different. Anomalies of lesser geographic extent, called surface anomalies, are irregularities in the field caused by deposits of ferromagnetic materials in the crust. Such deposits contribute little to the main field except in localized areas. The slow secular change of the main field is then attributed to changes in the strength and distribution of the internal current system as well as to the westward drift.

4.3 Geomagnetic Coordinates

The errors in describing the magnetic field at the Earth's surface by means of the centered dipole method of section 4.1 can be as great as 30% in some locations. This error can be reduced to about 10% by displacing the dipole axis about 400 km toward the western Pacific from the center of the Earth. Increased accuracy requires a more sophisticated mathematical representation, such as a multipole expansion of the field. The higher order terms in the multipole expansion fall off quickly with increasing distance from the Earth, and at several Earth radii it is generally accurate enough to use the dipole-field description.

A widely used method to describe the surface magnetic field is to fit the worldwide set of field measurements by a spherical harmonic analysis. This fit gives a surface field accuracy of roughly 0.5%. It is possible to develop an improved set of coordinates (called "corrected geomagnetic" coordinates) by using the difference between the dipole and spherical harmonic models.

Computation of the corrected geomagnetic coordinates begins by starting in the equatorial plane at the same point with a dipole field line and a spherical analysis field line, and then calculating the distance between the "landing points" of the two field lines on the Earth. In its simplest form, the method consists of labeling the spherical analysis field lines (sometimes called the "real field" lines) with the coordinates of the coincident equatorial, dipolar field lines. The spherical analysis field has numerous irregularities due to regional anomalies and the geometric asymmetries and so it is difficult to assign a meaningful symmetric grid pattern to such a system. However, superimposing the symmetric dipolar grid system on the "realistic" spherical analysis produces a useful coordinate system for modeling purposes.

An additional coordinate parameter which has been effective in organizing the behavior of geomagnetically trapped plasma is the invariant latitude, Ψ. This parameter uses the BL coordinate systems, where B is the magnetic field strength and L is a magnetic shell parameter. The L-shell is the surface traced out by the guiding center of a trapped particle as it drifts in longitude about the Earth while oscillating between the mirror points. Along most field lines L varies by less than 1%. On a given L-shell a particle's physical distance from the Earth may change as it circles the Earth because of the asymmetries in the geomagnetic field (especially at high altitudes where the distortions due to the solar wind interactions become important). At very high altitudes (where particle trapping is not possible) the L-shell method is no longer applicable. For a pure dipole field, L is equivalent to the shell's equatorial radius. That is, $L = R/R_E$ where the field lines intersect the geomagnetic equitorial plane at a distance R from the center of the Earth, and R_E is the radius of the Earth. In terms of L-shells the invariant latitude is defined as

$$\psi = \cos^{-1} \sqrt{\frac{1}{L}}. \qquad (4.16)$$

Another set of coordinates is

$$R = L \cos^2 d \qquad (4.17)$$

and

$$B = \frac{M}{R^3} \sqrt{4 - \frac{3R}{L}} \qquad (4.18)$$

where d is the magnetic dip latitude. In order to explain dip latitude we need to digress a moment and return to the geomagnetic field geometry.

Earlier we described the geomagnetic poles as the intersection of axis of the best fit centered dipole with the surface of the Earth. However, the "magnetic poles" often indicated on charts and maps are not the geomagnetic poles. Instead the magnetic poles are the positions where the magnetic field becomes vertical (the dip angle I, is 90°). These magnetic (dip) poles are located asymmetrically at the geographic coordinates about 75°N; 259°E and 67°S, 143°E for epoch 1960. The magnetic dip equator is the locus of points where the magnetic field is parallel to the Earth's surface (I = 0°). The magnetic dip latitude (or dip latitude) is given by

$$d = \arctan(\tfrac{1}{2} \tan I).$$

Other useful geomagnetic parameters are geomagnetic longitude and geomagnetic time. The prime geomagnetic meridian is the Greenwich (England) half of the great circle that passes through both the geomagnetic and geographic poles, and geomagnetic longitude is measured eastward from this prime geomagnetic meridian. Geomagnetic time is defined analogous to conventional time by replacing the geographic axis with the geomagnetic axis and defining local noon as being the time when the Sun crosses the local geomagnetic meridian.

4.4 Daily Magnetic Variations (Quiet Days)

Ground based magnetic measurements show a repetitive diurnal variation on geomagnetically quiet days. The variations occur mostly during the day, and nighttime variations are, at most, of very small amplitudes. Coordinated worldwide studies show that these variations are global. Since the variation is controlled by the Sun and occurs on magnetically quiet days, the variation is naturally called the solar quiet daily variation or simply Sq. The source of Sq is a system of electric currents in the lower ionosphere produced by solar heating (and atmospheric tides caused by the Sun) which in turn drives currents in the highly conducting lower ionosphere (E-region).

Another source of variation is the tidal oscillations of the atmosphere due to the gravitational force of the Moon.

The resulting electric current produces the lunar geomagnetic variation which is denoted by L. The amplitude of L is about one tenth of Sq; typical ranges in the magnitude of Sq at the earth's surface are roughly 50 to 250 γ at the dip equator and 30 to 60 γ at magnetic latitudes of 10 to 60°—but Sq can vary significantly from day to day, with season and solar activity. The idealized, equivalent current system which reproduces the Sq variations is shown in Figure 4.3. The pattern of the current remains stationary with respect to the Sun, and the Earth rotates under it once a day from left to right in the diagram. The current vortex (centered about 30° dipole latitude) flows in a counterclockwise direction in the Northern Hemisphere and in a clockwise direction in the Southern Hemisphere.

The concentrated ionospheric current along the magnetic dip equator is called the equatorial electrojet and its existence can be predicted from the following argument. The first step is to align our coordinate system following the normal convention with the z axis parallel to the magnetic field. Thus, at the magnetic equator z is northward, x is along the local vertical, and y is aligned eastward along a tangent to the magnetic equator.

At the geomagnetic equator, the electric field is generally oriented east-west and the ionosphere can be considered to be horizontally stratified. Therefore, a very good approximation is to assume there are no vertical currents flowing through the equatorial ionosphere into the plasmasphere and above.

Returning to section 1.2, we apply equation 1.7 and the definition of the conductivity tensor (σ):

$$J_x = \sigma_p E_x + \sigma_H E_y \qquad (4.19)$$

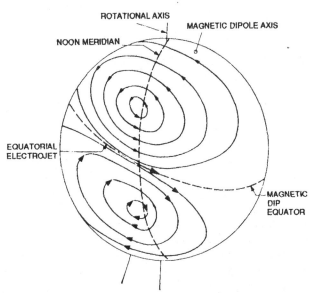

Figure 4.3 Idealized equivalent current system for the solar daily variation, Sq (*after Suguera and Heppner, 1968*).

$$J_y = \sigma_p E_y - \sigma_H E_x \qquad (4.20)$$

$$J_z = \sigma_{\parallel} E_z \qquad (4.21)$$

Since the vertical current (J_x) is zero, we find

$$J_y = \left(\sigma_p + \frac{\sigma_H^2}{\sigma_p} \right) E_y. \qquad (4.22)$$

In equation 4.22 the term

$$\left(\sigma_p + \frac{\sigma_H^2}{\sigma_p} \right)$$

is called the Cowling conductivity, and represents an area of enhanced conductivity due to the formation of a vertical polarization electric field. Typical observed values of the Cowling conductivity are sufficient to account for the intensity of the equatorial electrojet near 100 km altitude. In addition, horizontal electric field strengths are about 0.5 millivolt/meter and vertical polarization field strengths are approximately 10 mV/m. At the geomagnetic equator, the Pedersen conductivity has a sharp peak at an altitude of about 150 km with a value of about 3×10^4 mho/m, and falls off almost linearly to zero near 100 km and 220 km. The Hall conductivity has a very broad peak near 130 km with a value of about 4×10^4 mho/m and falls off rapidly to zero near 150 km and 90 km. The resulting Cowling conductivity has a very narrow peak near 100 km with a value near 8×10^3 mho/m. The Cowling conductivity falls off very quickly below the peak to near zero at 85 km; above the peak, the conductivity falls by one third at 120 km and decreases slowly to zero near 160 km.

The equatorial electrojet has been observed to rapidly change its direction in both geomagnetically quiet and disturbed times. The rapid reversals during disturbed periods are attributed to magnetospherically driven changes in the ionosphere. Quiet time reversals are most likely due to lunar driven atmospheric tides.

4.5 High-Latitude Magnetic Disturbances

The high-latitude geomagnetic field experiences more frequent and severe disturbances than do the middle- and low-latitude regions. The disturbances are statistically greatest in the auroral ovals, a narrow oval shaped belt encircling dipole poles in each hemisphere (see Figure 4.4). The most obvious characteristic of the oval is the visible light emissions (aurora) that are used to demarcate its boundaries. The equatorward edge of the oval marks the boundary between the magnetic field lines that are essentially dipolar, and the highly distorted field lines which extend far into the magnetotail (see Chapter 5). The auroral region is thickest in latitude in the nighttime sector ($7°$ or more), whereas during the day, the aurora becomes weaker, with less latitudinal extent. Because the auroral zone is a direct link with the geomagnetic tail, solar wind interactions which disrupt the tail are seen shortly thereafter as enhanced auroral activity.

Magnetospheric disturbances produce an equivalent ionospheric current distribution very similar to Figure 4.5. In this figure, current lines may not necessarily be continuous because portions of the polar ionospheric current systems may be completed within the magnetosphere (see Chapter 5). The major features of the high-latitude current system are: (1) the auroral electrojet, an intense westward flowing current stretching from the early morning sector to just beyond midnight within the auroral belt; (2) a weaker, eastward flowing auroral electrojet extending from the evening sector to just before midnight; (3) a general dusk to dawn current flow across the polar cap; and (4) a weaker current system spreading outward to lower latitudes. The boundary between the eastward and westward electrojets is called the Harang discontinuity. The total current in the auroral electrojet is about 10^6 A and is driven by the electrical coupling between the high-latitude ionosphere and magnetosphere. Although shown for only one hemisphere, approximately the same distribution appears to form simultaneously in the other hemisphere.

Enhancements of the polar ionospheric current system reduce the magnitude of the surface magnetic field strength. Variations in the H component can vary between 100 to 2000 γ beneath the auroral electrojet. The magnetic indices K_p and AE (see section 4.7) are used as measures of the intensity of magnetic disturbances in the auroral region.

4.6 Magnetic Storms

Worldwide magnetic disturbances lasting one or more days are called magnetic storms. These disturbances tend to occur simultaneously around the Earth but, on occasion, there are moderate disturbances in the auroral zones (auroral substorms) that are hardly noticeable at low latitudes. At a given location, the intensity of the disturbance is more a function of local time than of universal time.

Usually magnetic storms begin with a sudden worldwide increase in the magnetic field. This so-called sudden commencement (SC), appears as a rapid increase on H over a 1 to 6 minute interval and appears nearly simultaneously (within a minute) worldwide as an increase of about 5–30 gammas (see Figure 4.6). Weak storms tend to have a less pronounced and prolonged SC peak.

Within one-half to a few hours after SC, measurements of the surface magnetic field begin to show a worldwide decrease. This main phase of the storm lasts about one to three days, during which H slowly returns to the pre-storm value. However, many large random variations can appear during the main phase. The field decrease, often called a *Dst* decrease, is due to the development of a ring current at about three to six Earth radii. Observations show that injected

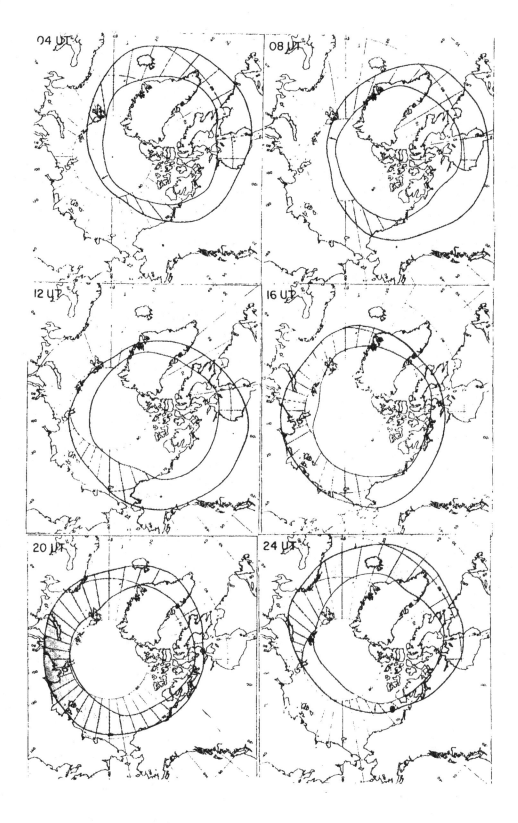

Figure 4.4   Approximate location of the auroral oval in the Northern Hemisphere at different UT hours. In the Southern Hemisphere a similar configuration occurs simultaneously *(after Akasofu, 1968)*.

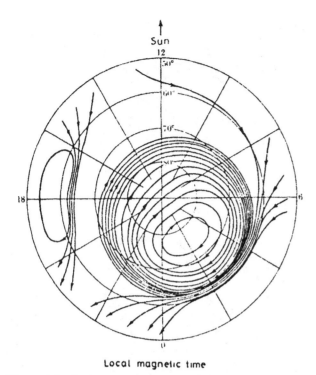

Local magnetic time

Figure 4.5 Equivalent current systems of a magnetic substorm. The concentration of current lines represents the auroral electrojects *(after Akasofu, 1968).*

magnetotail particles (10–100 keV range) are responsible for the ring current enhancements. The most intense activity tends to occur during the years near sunspot maximum.

The largest disturbances occur in the auroral zones and field deviations of about 2500 γ are not unusual. The amplitude decreases at higher latitudes to about one-half the auroral zone value near the geomagnetic poles. Equatorward of the auroral zone the disturbance intensity drops to about

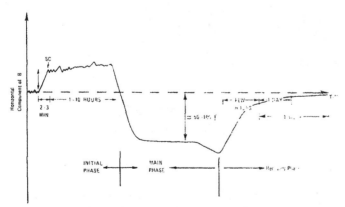

Figure 4.6 Example of magnetic field measurements during a magnetospheric storm. The baseline (horizontal axis) represents mean magnetospheric conditions. The symbol SC is used to denote the start of the sudden commencement which typically has an intensity of 5–30 gammas (γ) *(after Prochaska, 1980).*

one-fifth the auroral zone maximum at 50° geomagnetic latitude and a broad minimum (⅛ maximum) extends below 30° geomagnetic. However, near the equatorial electrojet the disturbance intensity increases to about one-fourth the maximum.

Magnetic records show that sometimes magnetic disturbances reappear with a 27-day period (average rotation rate of the Sun). Sometimes these repetitive disturbances can be attributed to long-lived active regions on the Sun. But at times, magnetic storms occur even when the Sun doesn't have any visible active regions. Historically, the magnetic activity was associated with a "mystery" or M region on the Sun. Today, we believe this activity is associated with coronal holes (section 3.5) which are only "visible" in x-ray photographs of the Sun.

4.7 Magnetic Indices

The intensity of a magnetic disturbance during each Greenwich day is specified by a variety of indices. Differences in the character and intensity of the magnetic variations with latitude influence the choice and derivation of the magnetic activity indices. For example, in the auroral zone, time scales of an hour or less are important. Whereas in middle latitudes, indices for intervals of 3 hours to one day are most appropriate. Near the equator, indices based on an interval of a day or more appear most suitable.

The *K index* is a 3 hour interval measure of the irregular variations of standard magnetic field measurements (i.e., magnetograms) and is used as an indicator of the *general level* of magnetic activity caused by the solar wind. This index is quasi-logarithmic and the scale is reported as integers ranging from 0 through 9. The K index from Fredericksburg, Virginia, has been designated as the standard measure of geomagnetic activity for the continental United States.

The K_p index ("p" for planetary) is based on the K indices from twelve preselected (worldwide) stations between geomagnetic latitudes 48° and 63°. Values of K for a given station are first used to compute the K_s index ("s" for standardized) which attempts to filter out local and seasonal variations. The K_s index also ranges from 0 to 9 (9 being the most disturbed), but is broken into finer gradations and is quoted in thirds of an integer using the symbols, −, o, and +. For example, the interval from 3.5 to 4.5 is given by the K_s values 4−, 4_o, and 4+. Therefore, K_s can have 28 different values: 0_o, 0+, 1−, 1_o . . . . . 8+, 9−, and 9_o. Finally, the K_p value for each 3-hour interval is derived from the K_s values from the twelve selected stations.

Since the 3-hour K, K_s, and K_p indices are defined with a quasilogarithmic scale, they are not suitable for simple averaging to obtain a daily index. The a_K index is the conversion of the K index to a roughly linear scale, and similarly the a_p index is derived from the K_p index. The average values of a_K

and a_p over the eight 3-hour intervals in a day are defined as the daily single station index, A_K, and the average planetary index, A_p, respectively. Table 4.1 shows the correlation between K_p and A_p.

The *AE index* (auroral electrojet) is the principal magnetic index for auroral zone studies. This index is measured in minutes (either 2.5 or 1 minute intervals) from the H-component measurements from an array of auroral zone observatories. At each station, the maximum positive displacement of H defines AU while the maximum negative H displacement defines AL. The difference between AU and AL is AE and the average between AU and AL is called AO. In practice this means that at each station a quiet time level must be determined (function of local time) which is then subtracted from the H measurement, leaving only the perturbations at each site. Thus for each index time interval, the most positive perturbation is AU, and the most negative is AL for that time.

The global AE index is determined using the largest reported AU and AL values from all the reporting stations. Thus, AU usually comes from a station beneath the eastward electrojet (1200–2200 local geomagnetic time), whereas AL comes from a station beneath the westward electrojet, with the most frequent occurrence of AL from a site near 0300 local geomagnetic time. In one sense AU and AL are local indices because they are only the amplitude of the H perturbations at two sites. However, these local extremes are determined by comparison with all reporting auroral stations, and thus AE is, in a real sense, a global index.

The major problem with the AE index is the distribution of the reporting auroral stations. The stations were selected because of international cooperation and not by some master plan design. Thus the station distribution is far from optimal. The present network is concentrated within a latitudinal belt extending from 62.5° to 71.6° corrected geomagnetic latitude. During intense magnetic storms, the auroral zone expands equatorward and, at times, the auroral electrojet may be equatorward of all reporting stations. Furthermore, auroral substorms are often limited to only certain local times (limited longitudinal extent) and, therefore, may be missed in part or entirely by the AE network. Another important problem is that AU or AL can result from movement of the electrojet relative to the station and thus can be misinterpreted as an electrojet enhancement. But in spite of its problems, AE remains our best auroral index.

The *Dst index* is the most widely used low latitude index.

The index represents variations of the H component due to changes of the ring current and is computed at hourly intervals because of the overall network (only four) of recording stations.

Micropulsations are the observed rapid variations of the geomagnetic field with durations ranging from a few minutes down to about 0.1 sec. Amplitudes (order of a few gammas) are largest in the auroral zone and are closely correlated with local values of K. The different types of pulsations have been divided into two groups: regular (and continuous) and irregular pulsations, denoted by pc and pi respectively. These two groups are further subdivided by pulse durations. Regular types (pc) have five categories: pc 1 (0.2–5 sec), pc 2 (5–10 sec), pc 3 (10–45 sec), pc 4 (45–150 sec), and pc 5 (150–600 sec). The irregular pulsations have only two subgroupings: pi 1 (1–40 sec) and pi 2 (40–150 sec). Most people agree that these pulsations are associated with magentospheric activity, but the different generation mechanisms needed to explain the various types of micropulsations are still unresolved.

4.8 Trapped Radiation

Contrary to popular misunderstanding, the radiation belts do not contain radioactive nuclei nor do the belts shield us from external radiation. The name is a holdover from the first-ever Explorer I satellite experiments which were designed to study the distribution of cosmic ray radiation, but instead discovered a seemingly permanent concentration of high-energy charged particles high above the earth.

The Earth's magnetic field provides the mechanism which traps charged particles within specific regions, called the Van Allen belts, about the equator. The motion of the trapped particles follows the three adiabatic invariants discussed in section 1.10. Namely, gyromotion around the magnetic field with time scale of milliseconds, a latitudinal reflection between mirror points with a time scale of seconds, and a longitudinal drift around the Earth with time scales of tens of minutes. The trapping regions (both electrons and protons) extend from the geomagnetic equator to about ±50° geomagnetic, but the trapping altitude structure is not discrete. Instead, the trapped particles extend over a range of altitudes with areas of slightly higher average concentration defining the traditional radiation belts. The approximate radiation belt distributions for protons and electrons are shown in Figures 4.7 and 4.8 respectively.

Table 4.1 K_p correlation with A_p.

K_p	0_o	$0+$	$1-$	1_o	$1+$	$2-$	2_o	$2+$	$3-$	3_o	$3+$	$4-$	4_o	$4+$
A_p	0	2	3	4	5	6	7	9	12	15	18	22	27	32

K_p	$5-$	5_o	$5+$	$6-$	6_o	$6+$	$7-$	7_o	$7+$	$8-$	8_o	$8+$	$9-$	9_o
A_p	39	49	56	67	80	94	111	132	154	179	207	236	300	400

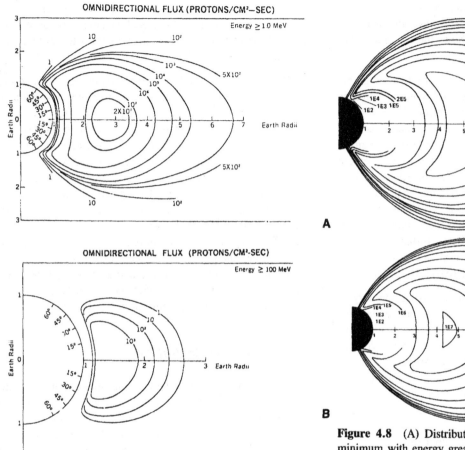

Figure 4.7   A, Distribution of trapped protons with energy greater than 1 MeV. B, Distribution of trapped protons with energy greater than 100 MeV *(after Smith and West, 1983).*

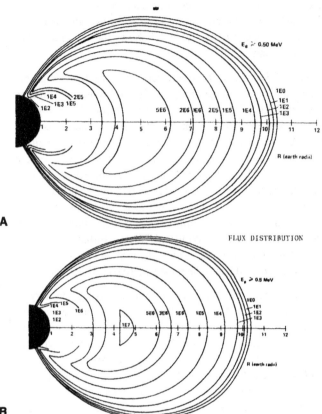

Figure 4.8   (A) Distribution of trapped electrons during solar minimum with energy greater than 0.5 MeV. (B) Distribution of electrons with same energy during solar maximum. The notation "E" refers to the power of ten: 1E4 = 1 × 10⁴ cm⁻²sec⁻¹, etc. *(after Smith and West, 1983).*

The radiation belts are approximately azimuthally symmetric, except near the South Atlantic anomaly. The magnetic field strength is lower than normal over the South Atlantic because of the dipole field geometry (see Figure 4.9), and therefore, the radiation belts reach their lowest altitudes in this area. Figure 4.10 depicts the inner radiation belt, showing how the South Atlantic anomaly extends down to lower altitudes.

Magnetic disturbances and solar cycle variations may modify the trapped radiation environment. Below geosynchronous altitudes, trapped protons trace out fairly symmetric orbits about the Earth. Local time effects are generally minor, although, factor-of-four variations in the background proton fluxes occasionally occur between noon and midnight. At low altitudes, the trapped protons are insensitive not only to local time effects but also to geomagnetic disturbances. For example, below 2 R_E (energies over 25 MeV) there is little response to even severe disturbances. At higher altitudes (1.0 MeV peak near 3 R_E) there is an observable response to even minor disturbances, with the decay to

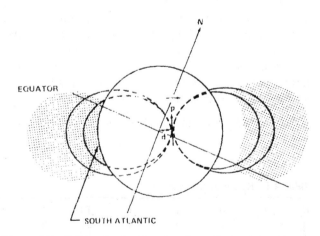

Figure 4.9   Schematic showing that the radiation belts (shaded area) are lower in the South Atlantic anomaly due to the offset of the dipole field *(after West et al., 1977).*

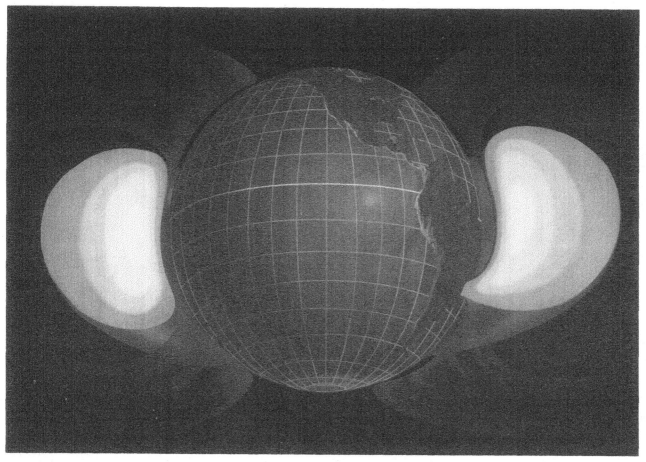

Figure 4.10 Illustration of the inner radiation belt showing the South Atlantic anomaly extending down to lower altitudes *(courtesy of Space Environment Laboratory).*

normal conditions requiring months or even years. Order of magnitude variations in the proton flux on time scales as short as 10 minutes have been observed above 5 R_E during geomagnetic storm conditions.

Near geosynchronous orbit, the trapped electrons (energies greater than 1 MeV) show a diurnal density variation such that the ratio between noon and midnight is about 3. These outer trapped electrons are also very sensitive to geomagnetic storms. Generally, the high energy electron flux may decrease (by factor-of-four) shortly after the onset of the geomagnetic storm, indicating that some kind of rearrangement is taking place. The high energy flux then increases to a maximum that may be ten to a thousand times as great as the pre-storm values. Subsequently, over the course of several days there is a gradual decrease to prestorm values. The source of these particles is still unresolved, and it is generally believed that they are *not* produced by the same physical mechanisms which cause the geomagnetic storm. In fact, some scientists have speculated that these particles originate in an extraterrestrial source, such as Jupiter, and arrive at Earth along IMF field lines which simultaneously intersect both planets.

The intensity of the trapped radiation which would be encountered in a 400 km polar orbit, and a 500 km 57° inclination orbit, is shown in Figures 4.11 and 4.12. The impact of such particles on both manned and unmanned space missions is discussed in Chapter 10.

4.9 Sources and Losses—Radiation Belts

Shortly after the detection of the radiation belt regions it was suggested that cosmic ray neutrons were the primary source mechanism for the inner belt. In this mechanism primary cosmic rays (protons with energies greater than 1 GeV) impinge on atomic nuclei in the high atmosphere and induce various nuclear reactions. A product of these reactions is a flux of neutrons leaving the atmosphere—the so-called albedo neutrons. However, a neutron has a short half-life (12 min), and many decay into a proton and an electron before leaving the magnetosphere. Some of these charged particles are trapped by the magnetic field and become members of a radiation belt. Today we believe that the process is probably

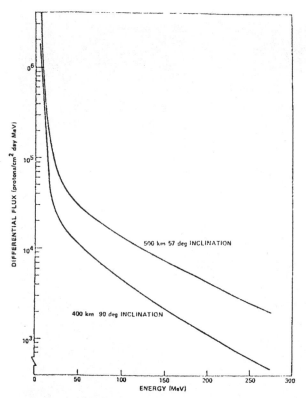

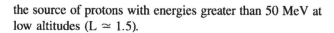

Figure 4.11 The radiation due to trapped protons encountered in a 400 km polar orbit and in a 500 km, 57° inclination orbit *(after Smith and West, 1983)*.

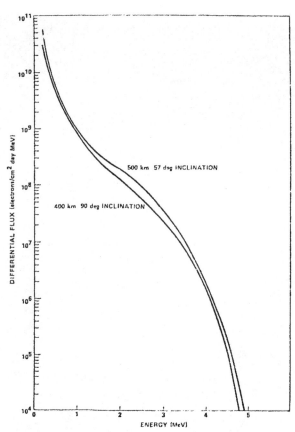

Figure 4.12 The radiation due to trapped electrons encountered in a 400 km polar orbit and in a 500 km, 57° inclination orbit *(after Smith and West, 1983)*.

the source of protons with energies greater than 50 MeV at low altitudes (L ≃ 1.5).

At lower energies, proton diffusion is thought to be the primary source mechanism. For example, suppose a ring of trapped particles (near the equator) drifts around the Earth on a constant magnetic field shell (second adiabatic invariant). However, during a geomagnetic disturbance the magnetic field lines are compressed more on the dayside than on the nightside and consequently, changes to the particle drift orbits smear the narrow ring into a broad band. Thus, there has been a transport of particles across L-shells. At higher altitudes (L > 1.8) diffusion becomes the dominating source. Furthermore, during geomagnetic storms, low energy protons can be injected into the higher altitude trapping region from deep in the magnetotail.

One reason why high energy protons are not found in the outer portion of the radiation belts is that the gyroradius for these particles is large enough to bring them into the lower region of the atmosphere. Before reaching their magnetic mirror points these particles are lost due to collisions with neutral atmospheric particles. Hence, energetic protons can remain trapped only in the parts of the Van Allen belt near Earth where the magnetic field is relatively strong (remem-

ber: gyroradius = $mv_\perp/qB$). At greater distances from the Earth, only protons of low energy (with relatively small gyroradii) will remained trapped. This restriction does not normally apply to electrons because their energies are rarely, if ever, large enough for the gyroradii to be more than a few kilometers, whereas proton gyroradii may be several hundred, or even thousand, kilometers.

The principal sources for the low energy particles in the outer belts are the solar wind and the ionosphere. The solar wind particles are introduced during geomagnetic storms, whereas the ionospheric ions are accelerated into the trapping regions by parallel electric fields (see section 5.7). During a magnetic disturbance, the low energy trapped electron number density increases rapidly due to injected magnetotail electrons, followed by an exponential decay in intensity. The decay is due to electron interactions which change their pitch angle and allow the electrons to penetrate deeper into the high latitude atmosphere where they are lost in collisions with neutrals. This process is called loss cone scattering (or pitch angle scattering), and therefore, the total electron intensity is then a balance between the injected tail electron and this loss process. In quiet magnetic periods, electron diffusion is an important electron source mechanism. Loss cone scattering also occurs with the trapped protons.

4.10 References

Akasofu, S-I. 1968. *Polar and Magnetospheric Substorms*. Springer-Verlag, New York.

Allen, J. H., and Feynman, J. 1979. Review of Selected Geomagnetic Activity Indices, in *Solar-Terrestrial Predictions Proceedings*, (R. F. Donnelly, ed.) NOAA/ERL, 7.

Busse, F. H. 1978. Magnetohydrodynamics of the Earth's Dynamo. *Annual Review of Fluid Mechanics*, *10*, p. 435.

Carrigan, C. R., and Gubbins, D. 1979. The Source of the Earth's Magnetic Field. *Scientific American*, *240*, Feb 1979, p. 118.

Gubbins, D. 1974. Theories of the Geomagnetic and Solar Dynamos. *Reviews of Geophysics and Space Physics*, *12*, p. 137.

Gubbins, D. 1981. Rotation of the Inner Core. *J. Geophys Res.*, *86*, p. 11695.

Hakura, Y. 1965. Tables and Maps of Geomagnetic Coordinates corrected by the Higher Order Spherical Harmonic Term. *Rep. Ionosph. Space Res. Japan*, *19*, p. 121.

Kamide, Y., and Akasofu, S-I. 1976. The Auroral Electrojet and Field-Aligned Current. *Planet Space Sci.*, *24*, p. 203.

Kamide, Y., Yasuhara, F., and Akasofu, S-I. 1976. A Model Current-System for the Magnetospheric Substorm. *Planet Space Sci.*, *24*, p. 215.

Kamide, Y., and Winningham, J. P. 1977. A Statistical Study of the Instantaneous Nightside Auroral Oval: The Equatorward Boundary of Electron Precipitation as Observed by the Isis 1 and 2 Satellites. *J. Geophys. Res.*, *82*, p. 5573.

Krause, F., and Radler, K. H. 1980. *Mean-Field Magnetohydrodynamics and Dynamo Theory*. Pergamon Press, Oxford, England.

McCrea, W. H. 1981. Long Time-Scale Fluctuations in the Evolution of the Earth. *Proc. Royal Soc. of London Ser A.*, *375*, p. 1.

Neshida, A. 1978. *Geomagnetic Diagnosis of the Magnetosphere*. Springer-Verlag, New York.

Prochaska, R. 1980. *Source Book of the Solar Geophysical Environment*. Air Force Global Weather Central, Offutt AFB, Neb.

Rich, F. J., and Basu, Su. 1985. Ionospheric Physics, Chapter 9 in *Handbook of Geophysics and the Space Environment*, ed. A. S. Jursa, Air Force Geophysics Laboratory, National Technical Information Services, Springfield, Va.

Roble, R. G. 1985. The Response of the High Latitude Thermosphere to Auroral Processes, AIAA/ Astrodynamics Conference, Paper No. AAS 85-316, Vail, Col.

Suguera, M., and Heppner, J. P. 1968. Electric and Magnetic Fields in the Earth's Environment, in *Introduction to Space Science*, Wilmot Ness and G. D. Mead, 2nd Ed., Gordon Branch, N.Y.

Smith, R. E., and West, G. S. 1983. *Space and Planetary Environment Criteria. Guidelines for Use in Space Vehicle Development, 1982 Revision (Volumes 1 and 2)*. NASA Technical Memorandum 82478.

Van Allen, J. A. 1991. Why the Radiation Belts Exist, *EOS*, *72*, No. 34, p. 361.

Vondrak, R. R., et al. (1979). Magnetosphere-Ionosphere Interactions, in *Solar-Terrestrial Predictions Proceedings*, R. F. Donnelly, ed. NOAA/ERL, *2*, p. 476.

West, G. S., Wright, J. J., and Euler, H. C. 1977. *Space and Planetary Environment Criteria. Guidelines for Use in Space Vehicle Development, 1977 Revision*. NASA Technical Memorandum 78119.

Williams, D. J. 1987. Ring Current and Radiation Belts, *Review of Geophysics*, *25*, no. 3, p. 570.

4.11 Problems

4-1. What is the difference between the Earth's surface magnetic field at the equator versus the poles? How well do these observations match a simple dipole model of the Earth's main field?

4-2. A polodial magnetic field is produced by zonal currents which flow along circles of constant latitude relative to the rotation axis. Whereas, electric mode magnetic fields are produced by currents which follow lines of constant longitude relative to the rotation axis. Finally, torroidal magnetic fields are produced by currents flowing parallel to the rotation axis. Into what category (or group of categories) does the Earth's main magnetic field belong?

4-3. What are the magnetic elements used to describe the Earth's main field? What is the difference between intrinsic and relative magnetic elements?

4-4. Why is the gravitationally powered dynamo a popular model for the Earth's main magnetic field?

4-5. What is the difference between geomagnetic and geographic coordinates? Is the difference significant? What changes are required to convert geomagnetic into corrected geomagnetic coordinates? Does the change make a significant improvement?

4-6. Distinguish between quiet time magnetic variations versus storm disturbances. Include such information as location, intensity, and duration.

4-7. What are the differences between the K_p, A_p, and AE magnetic indicies? Why do we need all of them? What are micropulsations?

4-8. Describe the particle distribution within the trapped radiation belts. What are the principle sources and losses (mechanisms) within these belts?

4-9. In section 4.4, we showed that the equatorial electrojet could be explained through the Cowling conductivity. Why is a vertically polarized electric field necessary? Is this vertically polarized electric field the source of the Cowling conductivity? Does a similar process apply to the equatorial electrojet or does some other mechanism apply? Explain your answer. (Hint: see section 5.7.)

Chapter 5

Magnetosphere

5.0 Introduction

In the absence of an interplanetary plasma, the Earth's dipole magnetic field would extend indefinitely in all directions. However, the geomagnetic field produces a semipermeable obstacle to the solar wind and the resulting interaction produces a cavity around which most of the plasma flows. The extent of this obstacle, called the magnetosphere, is related to changes in the solar wind density and velocity, and to variations in the strength and orientation of the interplanetary magnetic field (IMF). The magnetopause is the "surface" where the outward force of the compressed geomagnetic field combined with the magnetospheric plasma pressure, is balanced by the force of the solar wind plasma (see Figure 5.1). The bulk of the solar wind plasma is directed around the magnetosphere and does not approach the Earth any closer than the magnetopause. The resulting magnetosphere is shaped like a bullet, fairly blunt on the sunward side, and nearly cylindrical for a long distance in the antisolar direction. The stretched geomagnetic field lines forming the tail are bounded by a less distinct magnetopause than on the sunward side of the Earth. The magnetopause occurs approximately 10 Earth radii ($10\ R_E$) on the sunward side, and the magnetotail extends well beyond the orbit of the Moon (60 R_E). High-speed solar wind streams can compress the sunward side of the magnetopause to as little as 7 R_E.

As mentioned earlier, the solar wind acts as a supersonic (mach 8+) plasma in interplanetary space. Upon encountering the magnetosphere, a bow shock wave forms in front of the magnetopause. This shock is analogous to the aerodynamic shock found by a blunt obstacle in the supersonic flow of a wind tunnel, and the magnetopause corresponds to the "laminar" flow boundary behind the aerodynamic shock. The nose of the bow shock forms at about 15 R_E (see Figure 5.1), but its position is very sensitive to changes in solar wind conditions. The region between the magnetopause and bow shock is the magnetosheath and it is a region of rather disordered field, intermixed with heated, compressed, and irregularly distributed plasma.

The model of frictionless aerodynamic flow around an impenetrable obstacle fails to adequately explain the long, stretched cometlike "tail" of the magnetosphere. If the flow were really aerodynamic, then the magnetosphere should be teardrop shaped and have a relatively short magnetotail. Understanding the formation and dynamics of the magnetotail involves combining gasdynamics with electrodynamics, and the resulting magnetohydrodynamic (MHD) theory is complicated. For example, charged particles can move much more freely along the magnetic field than perpendicular to it, and consequently plasma properties such as electrical conductivity are highly anisotropic. The electric and magnetic fields in a plasma are determined by not only the local electrical charges and currents in the plasma but also by sources outside the plasma. These charges and currents are produced by plasma motions that are, in turn, determined at least in part by the forces associated with the electric and magnetic fields in the plasma. Therefore, it is not always possible to assume either the fields or their sources as "given" and to derive one from the other; one must derive the fields and their sources simultaneously and self-consistently.

5.1 Bow Shock

The bow shock is evidence of collective plasma behavior. As mentioned earlier, its general behavior is analogous to an aerodynamic shock about a blunt obstacle, but the analogy is limited. For example, theories about the detailed structure of the shock often involve heat transfer by various types of large-amplitude plasma waves. Unfortunately, none of these theories seems consistent with observations; it is particularly difficult to explain theoretically the great amount of ion heating that occurs.

Theorists expect that any sharp, well-defined shock wave in a magnetized plasma should obey certain "jump conditions" based only on the conservation of mass, momentum, energy, and magnetic flux. Mathematically these conditions are given by

$$[\rho v_n] = 0 \tag{5.1}$$

$$\left[\rho v_n \vec{v} + \left(p + \frac{B^2}{2\mu_o} \right)\hat{n} - \frac{1}{\mu_o} B_n \vec{B} \right] = 0 \tag{5.2}$$

$$\left[\rho v_n (\tfrac{1}{2} v^2 + h) + v_n \frac{B^2}{\mu_o} - \frac{1}{\mu_o} B_n \vec{v} \cdot \vec{B} \right] = 0 \tag{5.3}$$

$$[B_n] = 0 \tag{5.4}$$

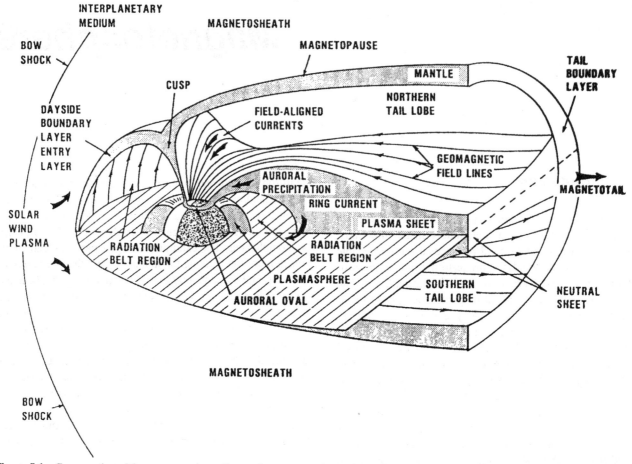

Figure 5.1 Cross section of the magnetosphere. For a quiet magnetosphere, geostationary altitudes are between the plasmasphere and plasma sheet (nighttime), and between the plasmasphere and dayside boundary layer (daytime). During active geomagnetic periods, geostationary satellites may become engulfed by the inward moving nighttime plasma sheet, and may pass through the daytime boundary (entry) layer (*after National Research Council, 1981*).

$$[E_t] = 0 \tag{5.5}$$

$$[B_n v_t - B_t v_n] = 0 \tag{5.6}$$

where the bracket notation is used to mean the value ahead of the shock is subtracted from the value inside the shock. That is,

$$[\rho v_n] = \rho v_n)_{\text{post shock}} - \rho v_n)_{\text{preshock}}. \tag{5.7}$$

In equations 5.1 through 5.6 the subscript n is used to mean "normal" (perpendicular) to the shock and the subscript t means "tangential" (parallel) to the shock. Equation 5.1 is a statement of conservation of mass across the shock boundary (or discontinuity). There is no discontinuity for the tangential component of the mass flow, and therefore there are no specific boundary conditions for the tangential component of the conservation of mass. Equation 5.2 is a statement of the conservation of momentum across the shock and equation 5.3 is a statement of the conservation of energy. In equation 5.3, h represents enthalpy per unit mass. Equation 5.4 results from the conservation of magnetic flux, and equation 5.5 follows from the fact that for steady state conditions, the divergence of the electrical field is zero [i.e.,

$\nabla \cdot \vec{E} = 0$]. Equation 5.6 follows from the frozen-in-flux criterion (i.e., $\vec{E} + \vec{v} \times \vec{B} = 0$). Taken together, equations 5.1 through 5.6 represent the boundary conditions which must be satisfied across any surface of discontinuity (or shock). For simplicity, we will use the subscript *1* for *preshock* values, and the subscript *2* for *postshock*. There are two important classes of solution for the above set of boundary conditions. The first type of solution is called "tangential" (B is tangent to the shock) where

$$v_{t_1} = 0 \tag{5.8}$$

$$B_{n_1} = 0 \tag{5.9}$$

and

$$\left[P + \frac{B^2}{2\mu_o} + \rho v_n^2 \right] = 0. \tag{5.10}$$

Thus, for tangential solutions there is no net flow of plasma along the shock surface, but there is flow across the surface.

In the second (or radial) type solution,

$$v_{n_1} \neq 0; \; v_{t_1} = 0 \tag{5.11}$$

and

$$B_{n_1} \neq 0. \qquad (5.12)$$

There are two possibilities for the postshock plasma:

$$V_{t_2} = 0 \text{ (fast shock)} \qquad (5.13)$$

or

$$v_{n_2} = \frac{B_{n_1}^2}{\mu_0 \rho_1 v_{n_1}} \text{ and } v_{t_2} \neq 0 \text{ (Alfven shock).} \qquad (5.14)$$

The fast shock results in a simple gas dynamic solution with

$$[B] = 0.$$

The B field does not affect the fast shock solution. However, this is not the case for the Alfven shock. For the Alfven shock,

$$\rho_2 = \frac{\mu_0 \rho_1^2 v_{n_1}^2}{B_{n_1}^2} \qquad (5.16)$$

and the velocities obey the following relationship

$$M_{s_1}^2 > \frac{M_{A_1}^2}{1 - \frac{(\gamma - 1)}{2}(M_{A_1}^2 - 1)} > M_{A1}^2 \qquad (5.17)$$

where

$$M_A = \text{Alfven mach number} = \frac{v}{v_A} \qquad (5.18)$$

and M_s is the gas dynamic mach number. In crossing the Alfven shock, the B field "bends" (i.e., acquires a tangential component), and since there is no electric field, the magnetic tension must be strong enough to direct the plasma flow along the "bent" field lines. The necessary condition for an Alfven shock is given by equation 5.17: the magnetic tension must exceed the plasma's dynamic pressure. The definition of the Alfven velocity, v_A, is given in equation 3.37.

How good are these above theoretical models? Satellite data shows that when the direction of the solar wind magnetic field is within about 40° (spiral angle) of the Earth's bow shock, the shock is thin and well defined and fits the tangential discontinuity model fairly well. However, when the solar wind magnetic field is more nearly perpendicular to the bow shock wave, the resulting shock structure is diffuse, turbulent, and ill-defined. It is then called a "pulsation shock," and unfortunately, the radial discontinuity theory is an inadequate description of this phenomena.

Using the boundary conditions for a tangential discontinuity one can show that the "jump" conditions across the shock are given by

$$\frac{v_{\text{postshock}}}{v_{\text{preshock}}} = \frac{\gamma - 1}{\gamma + 1} \qquad (5.19)$$

and

$$\frac{\rho_{\text{preshock}}}{\rho_{\text{postshock}}} = \frac{\gamma - 1}{\gamma - 1} \qquad (5.20)$$

where γ is equal to ratio of specific heats. Jump conditions shown in equations 5.19 and 5.20 are strictly valid in the limit of very high upstream mach numbers. For an ideal gas ($\gamma = \frac{5}{3}$) the velocity decreases by a factor of 4 and the density increases by a factor of 4 across the shock boundary. Although the solar wind is not an ideal gas, the above jump conditions are reasonable numbers for the Earth's bow shock. Computer stimulations show that the maximum jump conditions (equations 5.19 and 5.20) occur at the bow shock nose (region where solar wind flow is perpendicular to shock) and that the shock discontinuity weakens as one approaches the terminator. Equations 5.19 and 5.20 are analogous to a standard gasdynamic shock but we need to express a note of caution. The bow shock results from the complex interaction of both gases and electromagnetic fields; it is in fact a magnetohydrodynamic shock and it does have significant differences from gasdynamic shocks. In particular, the gas velocity parallel to the shock boundary is unchanged across the gasdynamic boundary. But the velocity parallel to a magnetohydrodynamic shock is usually changed by the shock because charged particles cannot freely cross magnetic field lines. Charged particles are constrained to spiral about magnetic field lines and the resulting electromagnetic interactions can increase the component of the particle's velocity parallel to the shock boundary.

Another difference is that the thickness of a gasdynamic shock is usually determined by the distance that particles can travel between collisions. Even though the electron gyroradius is several kilometers and the ion gyroradius is several hundred kilometers, the solar wind is so tenuous that the mean distance between particle collisions is on the order of 10^8 km (10^4 earth radii), far too large to be physically significant for the interaction between the solar wind and the magnetosphere. As a matter of comparison, the mean free path for air molecules near the surface of the Earth is about 10^{-4} m. In fact, the bow shock is not a collisional aerodynamic shock, but is due to the electromagnetic interaction between the magnetized solar wind and the highly conducting magnetosphere (region of intrinsic geomagnetic field). The aerodynamic calculations work reasonably well for the bow shock because the "frozen-in" interplanetary magnetic field limits the free particle motion of the solar wind plasma perpendicular to the field, and this constraint makes the collisionless solar wind plasma act like a collisionally dominated fluid, at least with regard to motion perpendicular to the field.

5.2 Collisionless Shocks: Wave Particle Interactions

A more complete analysis of collisionless plasma shocks shows the solution is more complicated than our simplified approach thus far because of the dispersive nature of plasmas. A plasma shock wave consists of many superimposed waves, or harmonics, of different wavelengths (Fourier's theorem). In fact, as the wave profile steepens, more harmonics of even shorter wavelengths are excited. Unlike waves in ordinary matter, the speed of plasma waves is frequency dependent.

This dispersive nature of plasma waves becomes important for waves propagating along a direction at an angle, but not exactly perpendicular, to B. Short wavelength waves propagate faster than do waves with longer wavelengths. The effects of plasma wave dispersion become more important as the shock front thins to about that of an ion gyroradius. At this time, the shorter wavelength components of the wave front move out ahead of the main shock, in effect widening the shock front by creating a weak leading edge. The length of the leading edge depends on how quickly the shorter wavelength waves can dissipate their energy. Other important plasma dispersion relationships are found in the limits when the normal to the shock is aligned perpendicular to the magnetic field, and when the normal to the shock surface is aligned parallel to the magnetic field.

First, we will examine the case of the shock normal aligned at right angles to the ambient magnetic field (i.e., tangential shock). In this case, the dispersion speed decreases with shorter wavelengths, and thus, short wavelengths trail behind the shock. Mathematically the dispersion relation is

$$\frac{\omega^2}{k^2} = v_p^2 = \left(\frac{\omega_{ci}}{k}\right)^2 + \frac{(\gamma_e kT_e + \gamma_i kT_i)}{M_i} \quad (5.21)$$

where v_p is the wave phase velocity, ω_{ci} is the ion cyclotron frequency, k is Boltzmann's constant, T_e is the electron temperature, T_i is the ion temperature, γ_e and γ_i are the specific heat ratios for the electrons and ions respectively, and M_i is the mass of the ions. Equation 5.21 describes an electrostatic ion cyclotron wave or ion acoustic wave. Ions can transmit an "acoustic-like" wave through the plasma by using the electric field to interconnect plasma particle motions. The second term in equation 5.21 is the sonic velocity of the plasma squared. The sonic velocity of the plasma depends on the electron temperature because the electric field resulting from the charge separation is proportional to the electron energy, and depends on the ion mass because the plasma inertia is principally due to the ions. Generally, electrons move so fast relative to the ion acoustic wave that the electrons have time to equalize their temperature and therefore, the electrons can usually be treated as isothermal (i.e., $\gamma_e = 1$).

As the ion acoustic wave propagates, electrons are pulled along with the ions, and the electrons tend to form a shielding layer around the bunched ions making up the waves. However, the compressed regions of the ion acoustic wave motion tend to widen and spread into the rarefaction regions for two reasons. First, the thermal motions of the ions in the compressed regions push the clumped ions into the adjacent rarefied areas, and second, the electric field formed by the surrounding electron cloud tends to pull apart the bunched ions.

A surprising property of a collisionless plasma is that wave energy can still be dissipated by a process called Landau damping, named after the famous Russian physicist, L. D. Landau, who first postulated the process could exist. Imagine a wave moving through a warm plasma. For charged particles nearly at rest, the propagating wave would appear as a rapidly fluctuating electric field. However, for those charged particles moving at a parallel speed nearly equal to the wave phase velocity, the wave appears virtually stationary and the particles "see" a constant electric field from the wave. These so-called resonant particles are accelerated by the wave's electric field, and thus, the particles take wave electric field energy and convert it into kinetic energy. This process is analogous to a surfer being propelled forward by "riding an ocean wave." Landau damping is an effective energy dissipation mechanism for ion-acoustic waves. As a result, the leading edge of the front does not propagate very far and the bow shock remains thin and well formed.

Ion acoustic waves are sometimes used to identify the presence of a solar wind induced shock. For example, project scientists for the Voyager satellite routed the measurements from the satellite's electric field probes through loud speakers to listen for the characteristic "sounds" of the ion-acoustic waves being reflected back upstream from the bow shocks expected to exist in front of both Jupiter and Saturn.

Similar plasma turbulence phenomena exist when the normal to the shock is parallel to the magnetic field (radial shock). The macroscopic behavior of such shock waves was discussed in section 5.1. In the case of parallel propagation, higher energy particles can stream ahead of the shock front because the magnetic field no longer retards their forward progress. For example, there is nothing to prevent overlapping particle streams from pushing ahead of the front. If the energy density of the plasma exceeds the energy density of the ambient plasma, then random motions in the plasma can flex the magnetic direction of motion; this instability is analogous to the oscillations which develop in a firehose when water rushes through the hose. In the case of the plasma, the magnetic field provides the "restoring" force which acts against the oscillations.

The oscillations formed by the firehose instability are hydromagnetic Alfven waves which propagate along the magnetic field at the Alfven velocity given by equation 3.37. For these oscillations, the Alfven velocity is the wave phase velocity, and the Alfven wave can exchange energy and momentum with resonant ions through wave-particle interactions. Theory predicts that the interactions act more like elastic collisions resulting in little change in energy. Therefore, Alfven wave turbulence inside the shock dissipates energy slowly resulting in a much thicker shock than for the case when the normal to the wave front is perpendicular to the magnetic field.

The discussion thus far shows the shock structure depends on the orientation of the shock relative to the geomagnetic field. Observations confirm the location and structure of the quasiparallel and quasiperpendicular portions of the shock front are always moving in conjunction with the changes in the IMF orientation. As mentioned in Section 5.1, measurements confirm the Earth's bow shock is thin when the normal to the shock front is nearly perpendicular to the geomagnetic field, and the shock is thick when the shock normal is parallel to the magnetic field. For example, satellite measurements in the early 1970s detected backward (i.e., sunward) fluxes of energetic particles, ion acoustic waves, and Alfven waves far upstream of the Earth's bow shock, consistent with quasiparallel and quasiperpendicular description presented above. Sometimes, these backwardly streaming particles and waves are referred to as the Earth's "foreshock."

5.3 Magnetosheath

Between the bow shock and the magnetopause lies the region of compressed and heated solar wind plasma called the magnetosheath. Figure 5.2 shows a schematic of the magnetosheath/magnetopause boundary. The top half of the figure shows a cross section along the noon-midnight meridian, and the lower half of the figure shows an equatorial slice through the same magnetosphere. Figure 5.3 also shows a magnetotail cross section which lies along the plane perpendicular to the cross sections shown in Figure 5.2.

Along the magnetopause boundary, the magnetosheath plasma interacts with the geomagnetic field. In general, the boundary layer plasma flow forms an electric field which is at some finite angle with the geomagnetic field. The orientation of the electric field can be approximated using equation 1.16 describing frozen-in flux conditions. We can define the

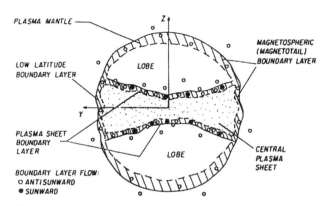

Figure 5.3 Schematic of magnetotail cross section in the y-z plane showing the major plasma regions in the magnetotail; the view is orthogonal to the schematic shown in Figure 5.2 (*after Alexander et al., 1985*).

magnetopause boundary condition in terms of available energy ($\vec{J} \cdot \vec{E}$) by using the differential forms of Ampere's law (equation 1.5)

$$\nabla \times \vec{B} = \mu_o \vec{J} + \mu_o \epsilon_o \frac{\partial \vec{E}}{\partial t} \qquad (5.22)$$

and Faraday's law (equation 1.6)

$$\nabla \times \vec{E} = -\frac{\partial \vec{B}}{\partial t}. \qquad (5.23)$$

Therefore, using equations 5.22 and 5.23 we can show that the available energy is

$$\vec{J} \cdot \vec{E} = -\frac{1}{\mu_o} \nabla \cdot (\vec{E} \times \vec{B}) - \frac{\partial}{\partial t}\left(\frac{\epsilon_o}{2}E^2 + \frac{B^2}{2\mu_o}\right). \quad (5.24)$$

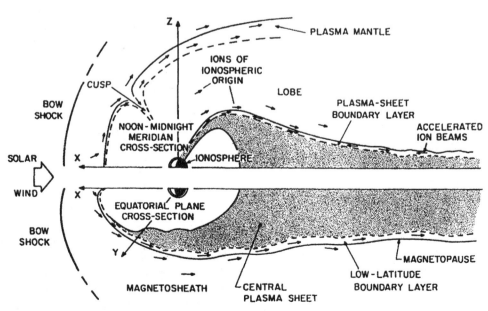

Figure 5.2 The top panel shows a schematic of a noon-midnight cross section (x-z plane) of the magnetosphere; the bottom panel represents an equatorial plane cross section (x-y plane) for the same magnetospheric conditions (*after Alexander et al., 1985*).

The first term on the right of equation 5.24 is the divergence of the electromagnetic Poynting flux which represents the energy flux carried by the electromagnetic field. The last term is the time rate of change of twice the energy density in the electromagnetic fields. In steady state, equation 5.24 shows that the rate of conversion of the electromagnetic energy into mechanical energy ($\vec{J} \cdot \vec{E}$) is equal to the divergence of the Poynting flux. Therefore, anywhere along the boundary where $\vec{J} \cdot \vec{E} > 0$, electromagnetic energy is converted into plasma energy, and wherever $\vec{J} \cdot \vec{E} < 0$, plasma energy is converted into electromagnetic energy. Satellite measurements show that magnetosheath plasma, momentum, and energy all generally decrease after crossing the magnetopause to form the plasma sheet boundary layer (see Figure 5.2). Thus, the boundary layer acts as an electromagnetic generator ($\vec{J} \cdot \vec{E} < 0$) converting plasma energy density into magnetic field energy. The corresponding energy dissipation region ($\vec{J} \cdot \vec{E} > 0$) is located in the tail current sheet.

5.4 Magnetospheric Currents

Thus far our analogy with an aerodynamic flow around an impenetrable object has been successful. However, the analogy breaks down when dealing with the cometlike "tail" of the magnetosphere (see Figure 5.4). The aerodynamic model predicts a relatively short, teardrop-shaped magnetosphere. Observations now indicate that tail-like plasmas (and magnetic-field configurations) have been seen occasionally at distances of more than 1000 R_E behind the Earth, but spacecraft coverage of the region beyond 60 R_E is extremely sparse.

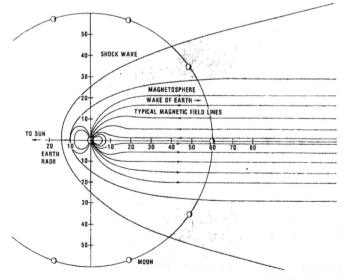

Figure 5.4 Cross section of the magnetosphere showing the approximate size in terms of number of Earth radii (1 Earth radius = 6,370 km). Geostationary orbits are at 6.6 Earth radii.

The existence of the long magnetospheric tail is probably the most remarkable feature of the magnetosphere. The cause of the magnetotail has been subject to some debate. One suggestion is that momentum is transferred from the solar wind to the outer part of the magnetosphere, similar to the frictional interaction that occurs when one fluid flows past another. The problem with this theory is to find the physical process equivalent to viscosity for the solar wind/magnetosphere interaction which is a collisionless process. Polarization electric fields and wave particle interactions are just two of the mechanisms invoked to explain this viscous-like process. Another suggestion is that the geomagnetic field lines can at times become connected to lines of the interplanetary magnetic field, and as the solar wind sweeps by the Earth, it carries these connected field lines on the downwind side of the Earth, creating the long magnetotail. However, there are long periods of time when the field connection is not expected to occur, and the geomagnetic tail is still observed to exist. Whatever the process, we know from elementary physics that the magnetotail (and compressed sunward geomagnetic field) is the result of the superposition of effects from a number of individual currents generated by the magnetohydrodynamic interaction of the solar wind with the geomagnetic field and plasma.

In simple terms, the total magnetic field can be written as

$$B_T = B_d + B_{rc} + B_{tail} + B_{magnteopause} \qquad (5.25)$$

where B_T is the total magnetospheric magnetic field, B_d is the contribution from the Earth's main field (see Chapter 4), B_{rc} is the magnetic field produced by the ring current, B_{tail} is the magnetic field contribution from the cross-tail current, and $B_{magnetopause}$ is the magnetic field originating from currents flowing on the magnetopause surface which confines the magnetic field contributions from the other three components within the magnetopause boundary (see Figure 5.5). Historically, the magnetopause currents are usually called the Chapman-Ferraro currents in honor of the two scientists who, in the 1930s, postulated the existence of some type of high-altitude current system to explain Earth based magnetometer measurements.

5.4.1 Magnetopause Currents

In section 5.3 we showed that the interaction between the magnetosheath plasma and the geomagnetic field generates a current boundary which defines the magnetopause.

This boundary is not perfect, observations show that about 0.5% of the mass and energy incident on the front of the magnetosphere and about 10% of the incident solar wind electric field leak through the magnetopause and cause a variety of dynamical processes within the magnetosphere. The magnetopause current can vary strongly with time because it is ultimately driven by the interaction with the highly variable interplanetary field. On average, this current is directed from dawn to dusk near the dawnside equatorial

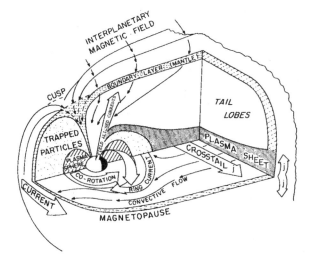

Figure 5.5 Cross section of the magnetosphere showing the principal current systems: magnetopause current, cross-tail (or neutral) current sheet, ring current, and field aligned currents. Also shown are the regions of convective and co-rotation plasma flow directions *(after Stern and Ness, 1981).*

plane, but short-term variations from this orientation frequently occur. Physically, the magnetospheric current must form a closed circuit. However, the exact path followed by this current is still unresolved.

5.4.2 Cross-Tail or Plasma Sheet Current

Another current system forms the neutral (or plasma) sheet which separates the oppositely directed magnetic fields emanating from the north and south polar caps. This current also flows from dawn to dusk and presumably merges with the tail current to form a closed circuit. The neutral sheet current amounts to about 10^5 amperes per R_E length out to about 30 R_E; beyond 30 R_E the current strength slowly decreases. The neutral sheet itself has average particle energies ranging from about 0.5–1 keV for electrons and 2–5 keV for ions, and the typical particle densities for both the electrons and protons is about 0.5 cm^{-3}. The undisturbed neutral sheet is about 5 R_E thick and it has an identifiable inner edge about 7 R_E from the Earth at local midnight. However, both the thickness and the position of the inner edge of the neutral sheet are variable.

Satellite measurements show that between 10 and 20 R_E the flux density is higher at the dawn and dusk flanks than at midnight, and the dawn field can be more than twice as large as that near dusk. In addition, the neutral sheet close to the Earth shows a seasonal warp. During the northern hemisphere summer, the neutral sheet rises above the solar magnetospheric plane at the center of the tail, and dips below the plane at the flanks. During the winter, the curvature is reversed.

Earthward of 100 R_E the bulk velocity in the plasma sheet has both earthward and tailward components when the

z-component of the IMF (B_z) is northward. When B_z is southward, observations of the bulk flow near 100 R_E show plasma sheet plasma moving predominantly tailward. Tailward of 180 R_E the bulk flow appears to be entirely directed away from the Earth at speeds near 500 km/sec.

Some researchers argue that the systematic tailward bulk flow when B_z is southward is evidence of tail-field reconnection forming earthward of 100 R_E. The standard growth phase model (see Section 5.9) has the reconnection occurring between 10 and 20 R_E. In this model the southward turning of the IMF enhances the magnetic reconnection on the dayside of the magnetosphere. As these newly connected lines are dragged into the tail (see Figure 5.17), magnetic and plasma energy are added to the tail, expanding the overall dimensions of the tail. Eventually the plasma sheet is severed relatively close to the Earth forming a plasmoid which moves tailward. The plasmoid expands adiabatically as it moves into the weaker tail-field, and observations show cross-tail expansion speeds near 80 km/sec while tailward speeds can range from 300 to 1000 km/sec.

A schematic showing the basic stages of the growth phase model is shown in Figure 5.6. Panel (a) shows the change in the auroral oval area during the substorm storage phase (approximately 40 minutes) which precedes the onset of the plasmoid expansion phase. Panel (b) shows the overall growth in the dimension of the magnetosphere during the storage phase by using the position of the ISEE-3 satellite as a reference point. Before the storage phase begins, ISEE-3 is clearly outside the plasma sheet. However, as plasma is added to the tail, ISEE-3 becomes engulfed by the plasma sheet. Panel (c) shows the plasmoid formation and movement tailward during the expansion phase of the storm. (Note: the redeployment of ISEE-3 from the L1 point needed a lunar gravitational assist to move it into the desired comet intercept course; the lunar maneuver kept ISEE-3 in the deep-tail for about 3 months in 1983.)

The plasmoid model is not the only explanation for the observed plasma behavior. Many researchers point out that the large bulk flows do not in themselves constitute proof of near-Earth reconnection. They point to many examples where streaming plasmas are observed without any discernable activity noticed in the polar cap or auroral zone. In addition, the magnetic signature usually taken as characteristic of plasmoid formation (B_z negative in the tail-field) can be interpreted in terms of local field-aligned currents. Section 5.9 presents a more detailed discussion of competing storm models.

5.4.3 Ring Current

A third current system, the ring current, encircles the magnetic equator, and is typically located between 3 to 6 R_E. Above 4 R_E the ring current involves a slow drift of particles across magnetic field lines, gradually moving them around

DE AURORAL IMAGES:

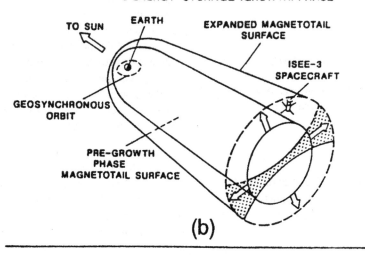

TAIL ENERGY STORAGE (GROWTH) PHASE

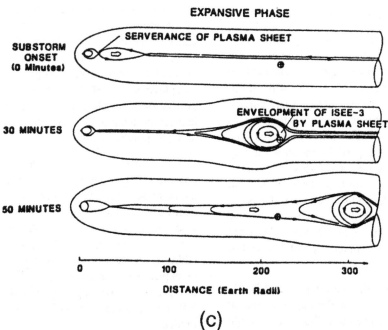

Figure 5.6 Top panel shows the change in area of the auroral oval during the growth phase of the magnetotail prior to substorm onset (oval sizes taken from Dynamic Explorer satellite imagery); panel (b) shows the growth of the tail during the same 40 minute period depicted in panel (a); panel (c) shows the subsequent sequence of events once the substorm begins (Note: panel (c) resets the time to zero—it is not the same as t_1 in panel (a)) *(after Baker et al., 1986).*

the Earth, westward for ions and eastward for electrons (see Figure 5.7). Typically, the particles making up the ring current have a mean energy of approximately 85 KeV and the majority (~90%) of the particles have energies ranging from 10 to 250 KeV. The ring current's total particle density and particle pressure peak around L shell values of 3 or 4 in the equatorial plane. Furthermore, the outer ring current magnetic field subtracts from the Earth's intrinsic dipolar main field within the area between the ring current and Earth's surface and adds to the nightside tail current. This reduction of the geomagnetic main field is most apparent during geomagnetic (or simply "magnetic") storms when the intensification of the ring current causes the magnetic field strength at the Earth's surface to drop, within hours, by up to 1 to 2%, with a gradual recovery taking about a day or two (see Figure 4.6). Observations show that during magnetic storms, injected plasma sheet particles are responsible for the ring current enhancements. Order of magnitude variations in the proton flux, on time scales as short as 10 minutes, can occur during geomagnetic storm conditions. Typical ring current density is about 10^{-8} Am^{-2}.

It was long believed that the ionic component of the ambient ring current consisted almost exclusively of solar wind protons, and it came as a great surprise when satellite measurements showed that the ring current had an appreciable O^+ component. Recent observations show that during geomagnetically active periods the percentage composition of ionospheric O^+ in the plasma sheet between 10 and 23 Earth radii can be as high as 40%. During periods of low geomagnetic activity the plasma sheet is dominated by H^+ and He^{++}. While He^{++} is clearly of solar wind origin, about 20 to 30% of the He^+ may be of ionospheric origin. A possible explanation is that a low-altitude electric field, parallel to the dipolar magnetic field lines, accelerates electrons downward into the ionosphere and accelerates positive ionospheric ions into the magnetosphere (see Figure 5.8). At high latitudes, the resulting outflow is called the polar wind. There is observational evidence that during disturbed geomagnetic conditions the parallel electric fields (see section 5.4.4) may be strong enough to produce a supersonic polar wind of H^+ and He^+ ions.

In addition, satellite observations show that probably most, if not all, O^+ and H^+ ions observed with energies less than 20 KeV originated in the ionosphere. These observations suggest the storm-time ring current forms in a two-step process. First, the outer trapping regions, beyond geostationary altitudes, are populated nearly continuously by a series of energization/injection events which generate the resident population consisting of a mixture of solar wind (normally high charge states, i.e., O^{6+} and He^{++}) and ionospheric (low charge states, i.e., O^+) ions. Second, during geomagnetic storms, the enhanced electric fields allow the outer ring

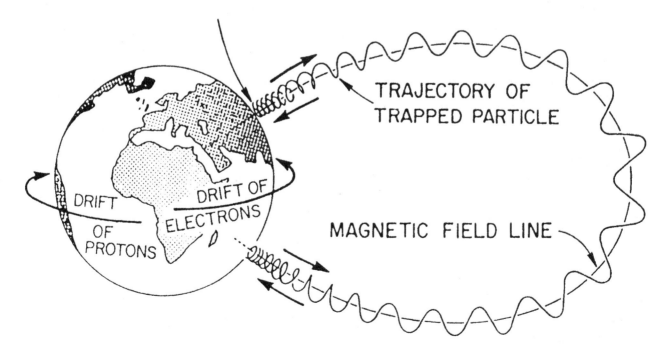

Figure 5.7 Motion of charged particles trapped by the Earth's magnetic field. The mirror point is the position where a charged particle stops and changes direction (i.e., a trapped particle bounces between its mirror points). Besides the two-degree motion of spiraling back-and-forth along the field line, the charged particles also have a third degree of motion, a drift around the earth. Notice that electrons drift eastward and protons drift westward, and the resulting charge separation produces a ring current system *(after Stern and Ness, 1981)*.

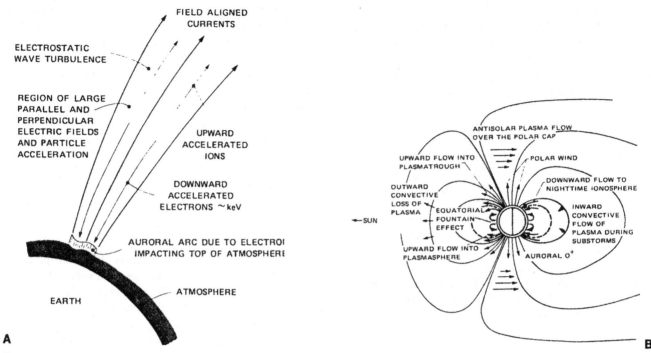

Figure 5.8   (A) Pictorial representation of the parallel electric fields that accelerate electrons downward and positive ionospheric ions upward into the magnetosphere *(after Space Science Board, 1978)*. (B) Transport processes that redistribute plasma in the ionosphere-magnetosphere system. Notice the O⁺ ions entering the magnetosphere from the high-latitude ionosphere *(after Burch, 1977)*.

current particles to adiabatically convect to lower altitudes, $(2 \leq L \leq 4)$ and form the storm-time ring current. As the storm subsides, the ring-current decays through pitch-angle diffusion scattering (see section 4.9) and charge exchange of ring-current ions in the hydrogen geocorona.

Thus far, our discussion of the outer ring-current attributed the ring-current to the drift of charged particles across the geomagnetic field lines. In section 1.7 we showed the general form of the drift velocity is given by

$$\vec{v}_d = \frac{\vec{F} \times \vec{B}}{eB^2} \qquad (5.26)$$

We also discussed three possible sources of drifts: electric field, magnetic field gradients, and the centripetal force on particles as they follow curved field lines. We also showed the electric field drift $(\vec{E} \times \vec{B})$ cannot produce a current because both the ions and electrons drift in the same direction, at the same rate. Therefore, the outer ring-current is formed by a combination of curvature and gradient drifts. However, at distances of 3 to 4 R_E an additional force, the particle pressure gradient, adds to, and can even dominate, the ring current formation. As mentioned earlier, the particle pressure peaks near 3 to 4 R_E such that above the peak the usual westward ring-current forms, but below the peak a weaker, but not negligible, eastward ring-current will form. In fact, the magnetic field variations measured at the surface of the Earth is a result of the ΔB decrease due to the westward current plus the ΔB increase from the eastward current.

Observations clearly show that the westward current dominates over the eastward current.

5.4.4   Field-Aligned Currents

Although it does not appreciably contribute to the overall geomagnetic configuration, the high-latitude field-aligned current system provides an important coupling mechanism between the magnetosphere and the ionosphere. Earlier in this century Birkeland concluded from his studies of auroral zone currents that these currents closed by vertical currents extending beyond the ionosphere. However, Sydney Chapman claimed that beyond the ionosphere the Earth was surrounded by a vacuum so there could not be any such vertical currents. It was not until the late 1960s and early 1970s that rocket observations definitely identified the field-aligned currents predicted by Birkeland, and many publications refer to these field-aligned currents as Birkeland currents.

There is an observed net flow of current into the ionosphere in the morning sector, and a net flow out in the evening sector. The currents vary with geomagnetic activity because the field lines along which the currents flow map back into the magnetotail. The total current fed into and out of the ionosphere by these currents ranges from 1 to 3 million amperes (typical current density is about $10^{-6} A/m^2$).

Detailed observations of the field-aligned currents show a two component structure. Figure 5.9 shows areas of down-

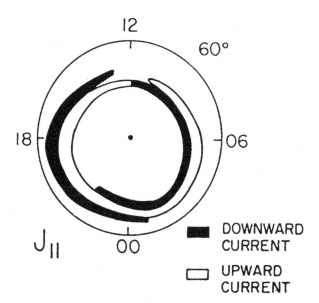

Figure 5.9 Schematic showing the typical distribution of Region 1 and Region 2 Birkeland currents during weakly disturbed geomagnetic conditions. Region 1 currents occur in the poleward portion of the auroral oval, and the equatorward portion of the field-aligned Birkeland currents are called Region 2 currents. Therefore, in the evening sector the Region 2 currents flow into the ionosphere and the Region 1 currents are flowing out of the ionosphere. The pattern is reversed in the morning sector. Polar cap field-aligned currents are not shown *(after Iijima and Potemera, 1978)*.

ward field-aligned currents (shaded) and upward currents (unshaded). The field-aligned currents in the poleward portion of the auroral oval are called Region 1 currents, and the equatorward portion of the field-aligned currents are called the Region 2 currents. Therefore, in the evening sector the Region 2 currents flow into the ionosphere and the Region 1 currents are flowing out of the ionosphere. The pattern is reversed in the morning sector (see Figure 5.14).

Observations show that the equatorward boundary of Region 2 is contiguous with the diffuse aurora in the post midnight sector, and extends about 2° equatorward of the electron precipitation boundary in the evening sector. There is also observational evidence of a field-aligned current system poleward of the Region 1 currents flowing along the open field lines forming the polar cleft (or cusp) (see section 5.5). These currents are sometimes referred to as the Region 0 currents and their intensity appears to be related to the B_y component of the IMF (see section 5.7).

Satellite measurements show the currents out of the ionosphere are carried mostly by precipitating electrons energized by upward directed electric fields forming parallel to the magnetic field lines. Similarly, the currents into the ionosphere are carried by electrons accelerated outward by downward directed field-aligned parallel electric fields. These parallel electric fields (sometimes called double-layers) are localized space charge regions which may build

up significant potential drops over distances of several Debye lengths (~10 km or more). For example, double-layers can form when the electron drift velocity equals the electron thermal velocity; at such times, an instability occurs which causes the ions and electrons to separate over short distances thereby producing short-lived electric fields parallel to B (see Figure 5.15). Other possible sources of double-layers are discussed in section 8.7. Measurements support the idea that a single double-layer is not responsible for the total observed particle energy, but instead many small double-layers together provide the total potential drop of 100 V to 10 keV. Theoretical calculations predict that double-layers are usually so thin that they are almost completely decoupled from either the upper ionosphere or the plasmasphere.

Within the ionosphere, observations show the electrons accelerated by parallel electric fields are associated with bands of higher energy electron precipitation. When the electron energy is plotted versus distance along the band, the spectrum resembles an "inverted V" structure, increasing from a few hundred eV at the edges to several keV at the core. Atmospheric Explorer data shows inverted-V structures are found throughout the region above 80° geomagnetic latitude. Below 80° geomagnetic, inverted-V events seem to concentrate in the late evening sector, and appear to be totally absent between 0900 and noon magnetic local time.

5.5 Magnetospheric Cusp or Cleft

Other important features of the magnetosphere shown in Figure 5.10 are the cusp (or cleft) and plasmasphere. The magnetospheric cusp is a narrow region, elongated in longitude and extending down from the high-latitude magnetopause into the polar ionosphere. It is filled with plasma nearly identical in properties to the magnetosheath plasma, and therefore it is generally assumed to originate there. The two cusps (one in each hemisphere) are in effect extensions of interplanetary plasma down to the Earth's upper atmosphere. It is usually taken for granted that the cusp consists of magnetic field lines that extend directly into the magnetosheath and the magnetosheath plasma, then enter the cusp simply by bulk flow parallel to the magnetic field. However, the region of open field lines contains both the cusp and the polar cap, and the question arises of why the entire polar cap is not filled with magnetosheath plasma. One possibility is that the polar cap magnetic field lines connect to the magnetosheath only very far downstream from the Earth, and the cusp field lines connect to the magnetosheath relatively nearby (see Figure 5.11). Another suggestion is that the lower latitude cusp field lines map into the dayside magnetosheath where the plasma flow is subsonic, whereas the polar cap field lines map into a region of supersonic magnetosheath plasma flow. Only the subsonic plasma flow satisfies the necessary flow constraints for entry into the lower magnetosphere and eventually down to the ionosphere. Whatever the origin of the cusp, it allows a sizable amount of low energy (on the order of a few hundred

Earth's Magnetosphere

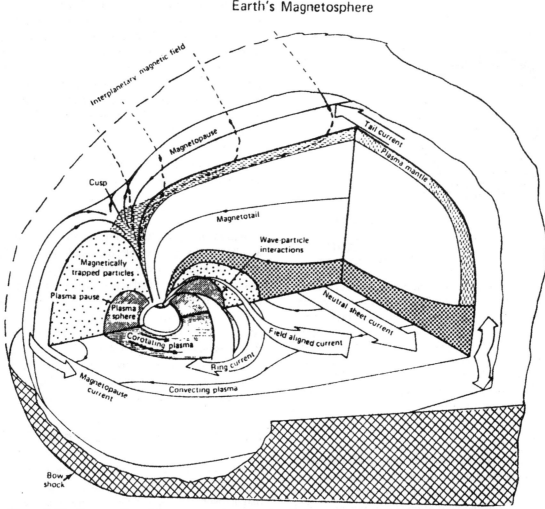

Figure 5.10   Cross section of the magnetosphere showing the important plasma regions and current systems *(after National Research Council, 1981).*

eV) plasma to precipitate into the ionosphere which, in turn, produces significant heating effects in the ionosphere, particularly during the polar winter when the ionization source from the Sun is absent.

In addition, observations from the Dynamic Explorer satellite showed that electric fields parallel to B allowed topside ionospheric ions to flow out of the polar cusp, and at higher altitudes these ions are dispersed across the polar cap by the convection electric field. Some scientists refer to this process as the cleft fountain effect analogous to the ionospheric equatorial fountain effect described in section 8.3. Some observational evidence indicates the cleft ion fountain, and not the broad polar cap ionosphere, is the principal source of the O^+ ions in the outer ring-current and plasma sheet.

5.6 *Plasmasphere*

The plasmasphere represents the relatively cold ionospheric plasma which is co-rotating with the earth due to frictional

coupling. The co-rotation can extend out to several Earth radii with particle densities ranging from about 10^4 cm^{-3} at low altitudes (1000 km) to 10–100 cm^{-3} near its outer edge. The boundary between the plasmasphere and the inner magnetosphere is called the plasmapause; plasma densities drop by at least a factor of 10 across this boundary. Typical plasmasphere energies are less than 1 electron volt. For the co-rotating plasmasphere, the plasma is stationary in its local frame of reference and therefore E = 0. However, to an observer on Earth, the plasmasphere plasma is kept in balance by an electric field which has to balance the Lorentz force or

$$\vec{E}_{rot} + \vec{v} \times \vec{B} = 0. \qquad (5.27)$$

Since the plasma velocity seen on Earth is merely the rotational velocity, equation 5.27 can be written as

$$\vec{E}_{rot} = -(\vec{\omega} \times \vec{R}) \times \vec{B} \qquad (5.28)$$

where ω is the angular frequency of co-rotation and B is the magnetic field strength at a radial distance R. The negative

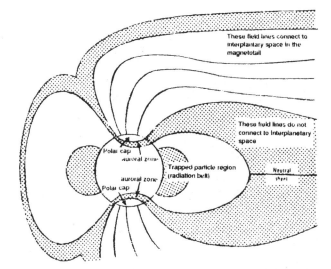

Figure 5.11 Sketch showing that the polar cap ionosphere has no conjugate field lines connecting the two polar ionospheres. Rather, these field lines extend far into the tail region and then connect with the interplanetary magnetic field. In this sketch the cusp field lines connect to the magnetosheath close to the Earth, thereby giving the cusp different plasma characteristics than the polar cap region. At lower geomagnetic latitudes, the magnetic field lines are basically dipolar (connecting both hemispheres). Notice that the auroral oval completely encircles the polar cap ionosphere *(after Townsend, 1982)*.

sign implies that the electric field is directed inward. In the non co-rotating outer magnetosphere, the cross-tail electric field is larger in magnitude than the plasmasphere's co-rotation electric field. It is generally accepted that the plasmapause represents the boundary between these two electric fields.

The location of the plasmapause can be estimated from the following expression (see problem 5-12):

$$E_{rot} = \frac{14}{L^2} \qquad (5.29)$$

where the units of E_{rot} are millivolts per meter (mV/m). Therefore, at $L = 4$, E_{rot} is about 1 mV/m which is comparable to the average strength of the cross-tail electric field. However, the instantaneous cross-tail electric field strength depends on the solar wind electric field which drives the cross-tail current. As expected, observations show the plasmapause position is a function of geomagnetic activity. Typically, at the geomagnetic equator under quiet geomagnetic conditions, the plasmapause geocentric height is between 5 and 6 R_E. During active geomagnetic conditions, the plasmapause geocentric height is approximately 3 R_E. The closest reported plasmapause height was about 2 R_E during an especially intense geomagnetic storm. As a general rule of thumb, the plasmapause lies approximately along the magnetic field line that maps down to the surface at 60° magnetic latitude.

5.7 Magnetospheric Convection

In the outer magnetosphere, the convection of plasma on the closed-field lines, including the neutral sheet, is generally directed toward the Sun (see Figures 5.12 and 5.13). The electric fields associated with convection are equivalent to the magnetospheric flow patterns and they generally map along magnetic field lines into the high-latitude ionosphere. Therefore, the magnetospheric plasma moves away from the Sun near the magnetosheath boundary layer and returns toward the Sun in the center of the magnetotail. As shown in Figure 5.14, the antisunward-sunward convection boundary maps along magnetic field lines into the ionosphere, driving field-aligned currents. From satellite magnetic field measurements, we know that a general field-aligned and ionospheric current system of this type exists at all times. However, the solar wind must somehow inject enough energy into the magnetosphere to sustain this convection system against the dissipative (joule heating) losses in the ionosphere. The power required to drive the steady state magnetospheric convection circuit has been estimated to be a few times 10^{10} watts. In order to maintain this power loss, about 10^{26} protons/sec (0.5% of solar wind particles) must enter the magnetosphere.

Figure 5.15 shows how the oppositely directed field-aligned currents flow along closed field lines down to lower

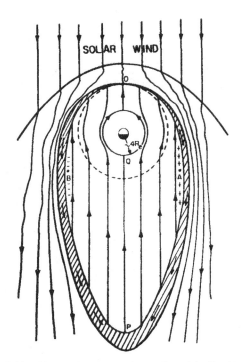

Figure 5.12 Sketch of the equatorial section of the Earth's magnetosphere looking from above the North Pole. Streamlines of the solar wind are shown outside the magnetopause boundary, and the interior streamlines depict the convective motion. The convective streamlines are also equipotentials of an associated electric field which may be regarded as being due to accumulations of positive and negative charges as indicated at A and B *(after Akasofu, 1968)*.

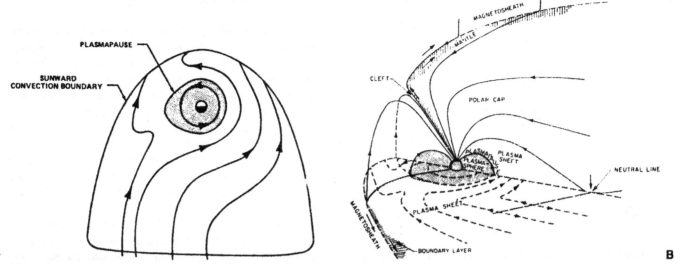

Figure 5.13 (A) Idealized plasma convection pattern in the equatoral plane of the magnetosphere, including the co-rotation within the plasmasphere. (B) Sketch of the magnetic field lines and various plasma regions of the magnetosphere. Only the northern dusk quadrant is shown for simplicity. Also drawn are contours denoting the convective flow of plasma in the equatorial plane. All magnetic field lines shown in this figure, except the two near the dusk meridian, lie completely in the noon-midnight meridian plane. The Van Allen radiation belts lie within and just beyond the plasmasphere *(after Burch, 1977).*

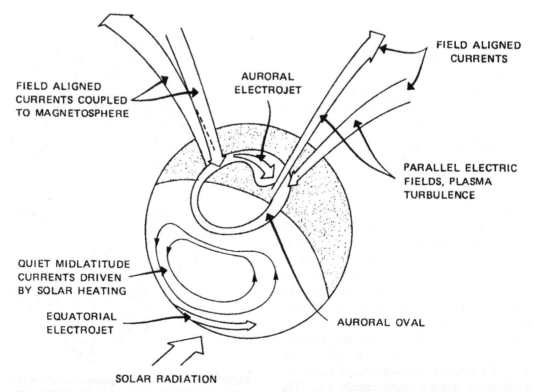

Figure 5.14 The configuration of the major current systems in the Earth's ionosphere. The connection to the magnetosphere is by the high latitude field-aligned currents. Within the ionosphere, the convection electric fields drive currents. In reference to Figure 5.9, Region 1 currents flow downward on the dawn side and upward on the dusk side of the poleward boundary of the auroral oval. Equatorward of the auroral oval, the field-aligned currents change direction (Region 2), upward on the dawn side and downward on the dusk side of the auroral oval *(after Space Science Board, 1978).*

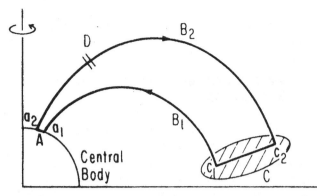

Figure 5.15 Sketch of the auroral circuit as seen from the Sun. B_1 and B_2 are dipolar magnetic field lines from the central body (Earth and ionosphere). C represents convecting plasma moving in the sunward direction (out of the figure). The interaction of moving plasma with the magnetic field induces an electric field which drives the field-aligned currents denoted by the arrows along B_1 and B_2. Also shown is a parallel electric field (or double layer) forming at D. Such a layer can form in a low density plasma because of the combined effects of electron thermal motion and electron drift motion. If the current density increases so that the drift velocity equals the thermal velocity, an instability occurs which causes the positive and negative charges to separate over short distances (i.e., resulting in a parallel electric field) *(after Alfvén, 1981)*.

latitudes. The flow of the ionospheric parts of these field-aligned current systems through regions of varying conductivity determines the detailed distribution of the electric fields in the auroral region (see section 8.5). There is some evidence that inside regions of auroral activity the convective flow intensifies and sometimes reverses due to variations of ionospheric conductivity.

The actual polar cap current pattern is sensitive to changes in the interplanetary magnetic field. In the northern hemisphere, a positive B_y component in the IMF produces mostly westward convection in the polar cusp region and a concentration of convection on the dawn side of the polar cap. As might be expected, a negative B_y component produces eastward convection in the cusp region and enhanced convection in the dusk sector of the polar cap. These B_y dependent effects are reversed in the southern hemisphere.

The Earth's magnetosphere is rarely quiet because the solar wind appears to continuously provide power to drive magnetospheric dynamics. For a southward B_z component of the IMF, the interconnection of the IMF with the geomagnetic field (see Figure 5.16B) allows the solar-wind electric field to map down into the polar cap resulting in a cross-polar cap potential drop of about 50 kV. Observations indicate potential drops exceeding 200 kV can occur during intense magnetospheric storms. When B_z is northward, the cross-polar cap potential does not drop to zero; observations seem to support a minimum potential drop of about 15 kV. This minimum potential implies a fraction of the solar-wind particles (and electric field) diffuse across the magnetopause,

even for northward B_z, probably due to wave-particle interactions as described in the next section.

5.8 Solar Wind Injection Into Magnetosphere

Three different mechanisms have been proposed for the transfer of solar wind particles from the magnetosheath into the magnetosphere.

1. Magnetic merging at the magnetopause
2. Diffusion of particles across the magnetopause
3. Magnetic-field drift of particles across the magnetopause

The first of these applies directly to the "open-magnetosphere" model while the other two are basically "closed-magnetosphere" model ideas, although they are readily adapted to an open-magnetosphere model as well.

In the closed-magnetosphere model, a closed boundary surface completely encloses the magnetosphere and separates it from the solar wind (see Figure 5.16). That is, no magnetic field line that goes through the Earth extends out into the solar wind. Conversely, the open-magnetosphere model has polar cap field lines extending out into the solar wind and joining with the interplanetary magnetic field. Observations appear to favor the open model, mainly because of the strong correlation between magnetospheric storms with the direction of the interplanetary magnetic

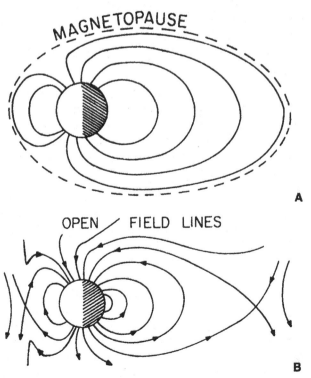

Figure 5.16 (A) Schematic view of a closed magnetospheric model and (B) view of open model. Both models are not to scale *(after Stern and Ness, 1981)*.

field. The open model would predict strong solar wind–magnetosphere interaction whenever the IMF has a southward component (with respect to the Earth) and this prediction has been verified by observations. However, theories fall short in accurately predicting the onset time of the ionospheric response to anticipated periods of dayside magnetopause merging. As a result, a number of models with very different response time scales have been proposed to explain the observations. Two of the most popular models are described in the next section.

Magnetic merging can happen in one of two ways: steady-state reconnection and localized sporadic reconnection. Figure 5.16B depicts the steady-state topology; this configuration allows solar-wind particles to penetrate the magnetopause by simple guiding center motion along open field lines. In addition, the magnetosheath particles have direct access to the polar cleft along the open field lines. In contrast, localized sporadic reconnection occurs only in limited regions for short periods of time. Such localized merging can produce small plasma tubes ($\sim 1\,R_E$ in size) which convect over the geomagnetic poles into the tail. Theories predict the tube-forming process twists the tube's internal magnetic field lines which helps preserve the tube's cylindrical shape. The magnetic "flux tubes" formed by this patchy reconnection process are sometimes called flux-transfer events (FTEs). Observations show FTEs form during southward B_z, but are almost never observed during B_z north conditions.

Turning to the second entry mechanism, particle diffusion is usually associated with wave-particle interactions and plasma instabilities forming along the magnetopause boundary. Under such conditions, the instabilities generate plasma waves capable of pitch-angle scattering particles across the magnetopause.

Similarly, the third mechanism, magnetic field drift, also relies on magnetopause instabilities. Sometimes, currents formed by the plasma motions associated with magnetopause instabilities generate short-lived, localized magnetic field structures which allow solar wind particles to drift across the magnetopause.

Regardless of the entry mechanism, most of the solar-wind particles which cross the magnetopause convect back to, and become temporarily stored in, the plasma sheet. However, there is an energy supply problem. Plasma sheet energies are in the keV range while solar wind particles start with only eV energies. Therefore, any plasma injection process must somehow accelerate the low energy solar wind particles up to the higher plasma sheet energies. A possible explanation is magnetic field annihilation. If the two oppositely directed tail lobe fields could come into contact, the resulting mutual annihilation would release about 5 keV worth of energy which might be transformed into plasma kinetic energy. The details of this possible energy conversion process are still unresolved, but there is little doubt that the particle energies can be accounted for in terms of the conversion of mag-

netic energy into particle energy (see the next section and section 8.7).

5.9 Magnetospheric Storms and Substorms

The level of magnetospheric activity varies widely. Variations in the solar wind produce time variations in the solar wind–magnetosphere interaction as the magnetosphere tries to adjust to a new steady state appropriate to the new solar wind conditions. Pressure variations in the solar wind often occur in the form of sharp discontinuities or shock waves. When these pressure discontinuities convect past the Earth, they often trigger large-scale, worldwide magnetospheric disturbances known traditionally as geomagnetic storms. As mentioned in Chapter 4, the main phase of the geomagnetic storm lasts from several hours to several days during which the worldwide surface magnetic field is generally depressed in magnitude because of impulsive injections of fresh plasma into the magnetospheric ring current.

This ring current enhancement is due to injected tail plasma in association with magnetospheric substorms. The short-lived substorm impulses are thought to be due to large-scale instabilities in the magnetospheric tail. As discussed earlier, the extended tail of the magnetosphere is maintained in part by a current that flows from dawn toward dusk in the neutral sheet. This current sheet is unstable to perturbations which allow magnetic merging to occur between 10 and 20 R_E and the subsequent acceleration of magnetospheric tail plasma (see Figure 5.17). Recent satellite observations indicate the excess free energy stored in the magnetotail can approach 10^{16} joules. Since a typical substorm takes about 40 minutes, a substorm can dissipate energy at a rate of about 5×10^{13} watts into the high-latitude ionosphere.

One possible model divides substorms into three phases: the growth phase, the expansion phase, and the recovery phase. During the growth phase, which lasts from a few minutes to an hour, the magnetic field strength in the tail increases in conjunction with the plasma sheet becoming gradually thinner in its north-south extent, shrinking toward the equatorial plane. The expansion phase begins with the "explosive" acceleration of the tail plasma, lasting several minutes or tens of minutes, into the inner magnetosphere enhancing the ring current. During the recovery phase the tail field relaxes to its prestorm configuration.

The above growth phase picture is most efficient when the solar wind has a large magnetic field component pointing southward. This observed correlation is easily explained in terms of the open magnetospheric model. A southward interplanetary magnetic field causes enhanced merging of geomagnetic and interplanetary field lines at the dayside magnetopause, setting up a temporary net transfer of magnetic flux from the dayside magnetosphere into the tail region (see Figure 5.17). The process continues until the merging

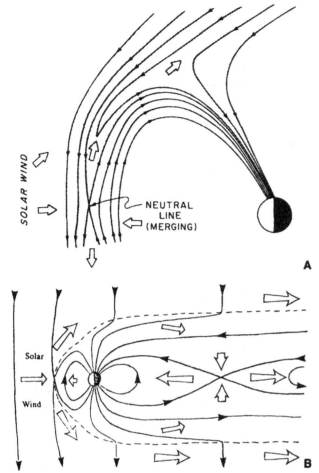

Figure 5.17 (A) Magnetic field topology in the interconnection ("merging") region of an open magnetosphere. Large arrows indicate plasma flow direction and arrows on field lines show field line direction. The view is in the noon-midnight meridian plane, with the Sun to the left *(after Hill and Wolf, 1977)*. (B) Cross section of the Earth's magnetosphere in the noon-meridian plane. The dashed line represents the magnetopause boundary. The large arrows show the convection system set up inside the magnetosphere by magnetic reconnection which occurs whenever the interplanetary field has a southerly component *(after Butler and Papadopoulous, 1984)*. The tailward moving plasmoid is shown in Figure 5.6.

rate in the tail adjusts itself to match the increased connection rate at the dayside magnetopause, so that the tail flux does not increase indefinitely. As mentioned above, the energy is stored in the increasing tail field for perhaps an hour or more before an instability occurs resulting in substorm onset. The substorm subsides when the reservoir of available magnetic energy in the tail is depleted, but a new substorm cycle can then start if the solar wind field remains southerly directed.

An alternative to the growth phase model is called the "directly driven" substorm model. This model is based primarily on the close correlation between substorm and solar wind conditions and on the premise that the state of the magnetosphere is determined solely by the solar wind. The correlation is described in terms of the epsilon parameter as

defined in section 3.7. Substorms should begin whenever epsilon exceeds a certain critical value which, in turn, triggers magnetospheric instabilities, and the substorm continues as long as epsilon remains above the critical value (see problem 5-10).

One important difference between the growth phase and direct models is the expected magnetospheric response to intervals of steady southward solar wind magnetic fields. The direct model would predict steady substorm-like conditions, while the growth phase model suggests that a sequence of substorms would occur, whose time scale would be determined by magnetospheric conditions. Another important difference concerns the expected behavior of the tail field at substorm onset. During onset, the direct model does not have the periodic dissipation of the increased tail energy as in the growth phase model. Rather, since epsilon will generally continue to increase as it passes the critical value, the tail field strength may also be expected to continue to increase. Therefore, in the directly driven model the tail field is expected to develop and decay in concert with epsilon. However, in the growth phase model the tail field is expected to grow before the substorm and generally to decay after the substorm onset.

5.10 Observations: Growth Phase Versus Directly Driven Model

Historically, statistical correlation studies compared solar wind density, speed, and magnetic field strength with 1 hour averages (or greater) of geomagnetic parameters such as AE, K, and Dst. However, such hourly averaging can mask important features of the solar wind–magnetosphere coupling because many solar wind parameters, and magnetospheric processes, occur on time scales much shorter than 1 hour. For example, the IMF's control of substorm onset time occurs on time scales finer than 1 hour. On the other hand, the variability of the solar wind bulk flow velocity, and its impact on the magnetosphere, changes slowly and hourly averages are sufficient.

Even when data is collected on the proper time scale, there is considerable disagreement in interpreting the data as to whether it supports the directly driven or growth phase model. For example, Figure 5.18 Shows two different interpretations of the same observations collected in September 1978. In Figure 5.18, the impulse response is analogous to a correlation coefficient; the larger the response fraction the more closely correlated are the input and output phenomenon. In both cases, the analysis used a method called linear prediction filtering. To understand how the same analysis technique can produce different results, we need to look at the linear prediction filtering methodology.

The linear prediction filtering method assumes the system under study can be represented as a linear, time-stationary, electrical system. In general

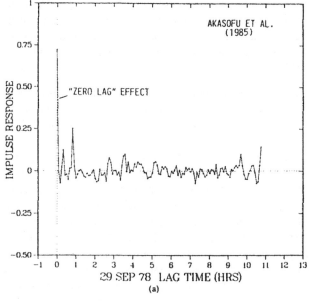

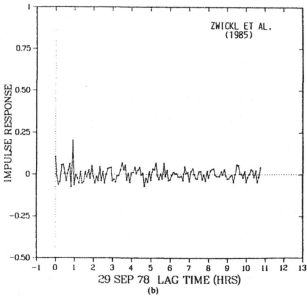

Figure 5.18 Results from the linear prediction filter method comparing the total output energy (UT) with E. Panel (a) assumes the ring current lifetime, τ_R (see equation 5.31) is a direct function of ϵ; as a result, the analysis shows a near zero lag time in the impulse response. Panel (b) uses the same methodology, but uses a constant ring current lifetime of about an hour; the resulting analysis shows a lag time of about 1 hour. In both panels, the impulse response is analogous to a cross-correlation function; large values of the impulse response correspond to a time delay between the solar wind input and the energy output *(after Baker et al., 1985).*

$$O(t) = g(t)I(t) \qquad (5.30)$$

where O(t) is the output time series representing the magnetosphere, I(t) is the input time series representing the solar wind, and g(t) is the filter which completely specifies the relationship between the solar wind and magnetosphere (wherever the two regions are linearly related). In Figure

5.18, the impulse response was derived from a linear prediction filter which compared ϵ (input) to the total energy dissipation in the magnetosphere (output). The total energy dissipation consists of the ring current energy injection rate (U_R), the joule heat production rate in the ionosphere (U_J), and the auroral particle energy flux (U_A). In practice, the total energy (U_T), in units of ergs/sec, can be approximated in terms of the Dst and AE indices. For example,

$$U_R = 4 \times 10^{10} \left(\frac{\partial Dst}{\partial t} + \frac{Dst}{\tau_R} \right) \qquad (5.31)$$

$$U_J = (2 \times 10^{15}) \times AE \qquad (5.32)$$

$$U_A = (1 \times 10^{15}) \times AE \qquad (5.33)$$

where Dst is usually corrected for dynamic pressure, AE is specified in nanoteslas, and τ_R is the ring current lifetime in units of hours. Even though the same basic equations were used in panels (a) and (b) of Figure 5.18, the computed response time differs because of the methodology used in computing τ_R. In panel (a), the directly driven model assumes τ_R is a function of ϵ and, as might be expected, the resulting linear filter prediction yields a near zero time lag between U_T and ϵ. In panel (b), τ_R was held constant (at 1 hour) for the analysis, and once again, it should then come as no surprise that the filter predicts a delay between U_T and ϵ.

This example is not meant to say one methodology is right and the other is wrong. Instead, the purpose of the example is to show the student to be cautious when interpreting any analysis and the conclusions drawn from that analysis.

Similar analyses have been completed for other input functions besides ϵ. The most popular alternatives to ϵ are VB_s and V^2B_s where B_s is the magnitude of the southward component of the IMF (sometimes written as $-B_z$). In fact, some authors argue that VB_s is a better predictor of AE than is ϵ during disturbed geomagnetic conditions; VB_s is the dawn-to-dusk component of the solar wind electric field. On the other hand, VB_s does not allow for any energy input when the IMF has a northward z component (i.e., $\theta < 90°$), whereas ϵ allows for a small energy input as long as $\theta > 0$. The lesson here is to remember there is no perfect predictor of solar wind–magnetosphere response; each input parameter has its usefulness if you recognize they each have limitations and biases.

Figure 5.19 shows the results of an analysis of a March 1979 storm using VB_s as the input parameter. At the time of the substorm the IMP-8 satellite was outside the bowshock and sunward of Earth. The two scales on the right side of the figure show the magnitude of VB_s ($-VB_z$) and the total available solar wind input energy (W_{in}) is measured in units of ergs/sec. The total ionospheric joule output energy was computed using a model of ionospheric currents derived from Earth based magnetometer measurements.

Notice a modest increase in the joule heat between 1020 and 1100 UT seems to track almost immediately with the

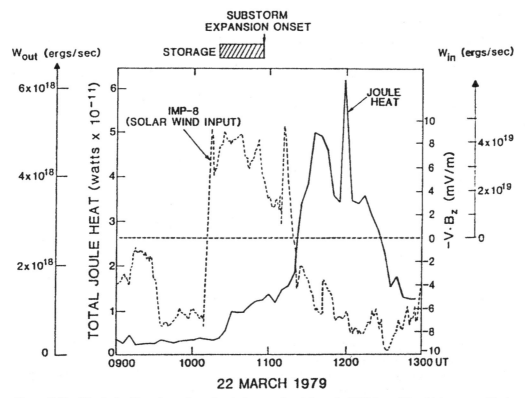

Figure 5.19   The dashed line shows the estimated energy input from the IMP-8 satellite which was outside the bow shock and sunward of the Earth. The solid line shows the computed substorm energy output (as joule heat) based on magnetometer measurements. The scale on the right side of the figure shows the estimated Energy input (ergs/sec) and the scale on the left side of the figure shows the computed total energy output (ergs/sec) as well as the total joule heat component to the total energy *(after Baker et al., 1985).*

increase of VB_s, as predicted by the directly driven model. Observations show that after 1100 UT intense auroral activity began, and the ionospheric current model generated a subsequent large increase in the joule heating rate which seems to favor the growth phase model. It appears Figure 5.19 supports the idea that even during the energy growth phase of the substorm, there is a relatively long interval of low energy dissipation by direct entry into the polar cap. Thus, the data suggests that both the directly driven and growth phase processes seem to be present.

Proponents of the directly driven model would argue that the results in Figure 5.19 are based on two (possibly) erroneous assumptions. First, the joule heating calculation uses a current system derived from magnetometer measurements; this computed current system may not be a unique solution—other current systems might exist which could give different joule heating results. Second, the joule heating term is not the only energy dissipation mechanism in the magnetosphere. If the other dissipation processes are included, Figure 5.19 might not have such a clear transition near 1100 UT.

Students should not be discouraged by these seemingly contradictory conclusions. This section points out that any analysis needs to be looked at with a critical eye. Basic facts about the magnetosphere come about through repeated ex-

periments showing different aspects of the very complicated nature of solar wind–magnetosphere coupling.

No matter which substorm mechanism is correct, the resulting atmospheric auroral display is spectacular. The auroral substorm begins with a sudden brightening of one of the quiet auroral arcs or a sudden formation of an arc, followed by a rapid poleward motion, resulting in an "auroral bulge" around the midnight sector. As the auroral substorm progresses, the bulge expands in all directions; the evening sector is characterized by a "westward traveling surge," and in the morning sector of the bulge, the arcs disintegrate into patches which drift eastward at speeds about 300 m/s. The expansive phase ends and the recovery phase starts when the expanded bulge begins to contract (see Figures 8.1 and 8.2).

The principal substorm magnetic signature is a decrease in the horizontal component of the surface geomagnetic field under the auroral bulge due to an embedded ionospheric (westward) "electrojet" current of about 10^6 amps. In addition, a spatially widespread magnetic signal usually appearing as a damped sinusoid (about 1 minute period) is initiated at the onset of expansion. This characteristic signal is known as a pi 2 and is thought to be excited by rapid changes in plasma flow in the magnetosphere. Another possible sequence has "multiple onsets," arising from observations of several discrete increases of activity during the auroral bulge expansion. Each of these increases is associated with a pi 2 burst, the formation of a new westward traveling surge in the auroras, and as the surge crosses the local meridian, a discrete poleward expansion of the westward electrojet. The first increase in activity is termed the "onset," and all subsequent increases are called "intensifications." More details on auroral substorms are given in section 8.5.

5.11   Summary

The solar wind is an ionized gas that streams away from the Sun. In passing the Earth, the solar wind interacts with the Earth's magnetic field and severely deforms it, carving out the magnetospheric cavity. A collisionless shock wave (the bow shock) forms upstream in the solar wind. Away from the Sun, the magnetosphere stretches out into a long tail—the magnetotail.

The magnetosphere is the region surrounding the Earth where the movement of plasmas is dominated by the geomagnetic field. The ionized gases in the magnetosphere take part in a gigantic pattern of magnetospheric circulation or "convection" in which the plasma over the polar regions is swept away from the Sun while plasma outside the auroral zone (i.e., at lower latitudes) is swept toward the Sun. At still lower latitudes (below about 60° geomagnetic) and altitudes, the plasma does not have this convection pattern but instead co-rotates with the Earth.

At times, the geomagnetic field is disturbed by the occurrence of magnetic storms which cause a small reduction in the geomagnetic field strength over most of the Earth's surface due to enhancement of the ring current. The substorm has been identified with large-scale changes in the magnetotail in which the total flux is suddenly reduced, thus producing large induced electric fields which drive tail plasma toward (and away from) the Earth. The most visible products of substorms are bright, active, and extensive auroras. It is fair to say that the basic elements of the magnetospheric storm (and substorm) are reasonably well understood but our theoretical understanding of how the solar wind triggers a substorm is still far from complete.

5.12   References

Akasofu, S-I. 1968. *Polar and Magnetospheric Substorms*. Springer-Verlag, New York.

Akasofu, S-I. 1980. Working Group Report on Geomagnetic Storms, in *Solar Terrestrial Predictions Proceedings*, R. F. Donnelly, ed. NOAA/ERL, vol. 4, p. A-91.

Akasofu, S-I. 1982. Interaction Between a Magnetized Plasma Flow and a Strongly Magnetized Celestial Body With an Ionized Atmosphere: Energetics of the Magnetosphere. *Annual Review of Astronomy and Astrophysics*, vol. 20, p. 117.

Akasofu, S.-I. 1985. Explosive Magnetic Reconnection: Puzzle to Be Solved as the Energy Supply Process for Magnetospheric Substorms?, *EOS*, vol. 66, p. 9.

Akasofu, S.-I., Olmstead, C., Smith, E. J., Tsurutani, B., Okida, R., and Baker, D. N., 1985. Solar Wind Variations and Geomagnetic Storms: A Study of Individual Storms Based on High Time Resolution ISEE-3 Data, *J. Geophys. Res.*, vol. 90, p. 325.

Alfven, H. 1981. *Cosmic Plasma*. D. Reidel Pub. Co., Boston.

Alexander, J. K., Bargatze, L. F., Burch, J. L., Eastman, T. E., Lyon, J. G., Scudder, J. D., Speiser, J. W., Voigt, G. H., and Wu, C.-C. 1984. Coupling of the Solar Wind to the Magnetosphere, in *Solar Terrestrial Physics: Present and Future*, D. M. Butler and K. Papadopoulas, eds., NASA Pub. 1120, Chapter 5.

Baker, D. N., Bargatze, L. F., and Zwickel, R. D. 1986. Magnetospheric Response to the IMF: Substorms, *J. Geomag. Geoelectr.*, vol. 38, p. 1047.

Baker, D. N., Fritz, T. A., McPherron, R. L., Fairfield, D. H., Kamide, Y., and Baumjohann, W. 1985. Magnetotail Energy Storage and Release During the CDAW-6 Substorm Analysis Intervals, *J. Geophys. Res.*, vol. 90, p. 1205.

Baker, D. N., Jones, E. W., Jr., Belian, R. D., and Higbee, P. R. 1982. Multiple-Spacecraft and Correlated Biometer Study of Magnetospheric Substorm Phenomena. *J. Geophys. Res.*, vol. 87, p. 6121.

Bauer, S. J. 1973. *Physics of Planetary Ionospheres*. Springer-Verlag, New York.

Berman, L., and Evans, J. C. 1977. *Exploring the Cosmos (2nd ed)*. Little, Brown and Co., Boston, Mass.

Block, L. P. 1973. The Magnetosphere, in *Cosmic Geophysics*, Egeland et al., eds. Universitetsforlaget, Oslo Sweden, Chapter 8.

Brandt, J. C. 1970. *Introduction to the Solar Wind*. W. H. Freeman and Company, San Francisco.

Burch, James L. 1977. The Magnetosphere, in the *Upper Atmosphere and Magnetosphere*. National Academy of Sciences, Washington, D.C., p. 42.

Burke, W. J. 1982. Magnetosphere-Ionosphere Coupling; Contribution from IMS Satellite Observation. *Rev. Geophys and Space Physics*, vol. 20, p. 685.

Burke, W. J., Hardy, D. A., and Vancour, R. P. 1985. Magnetospheric and High Latitude Ionospheric Electrodynamics, in *Handbook of Geophysics and the Space Environment*, A. S. Jursa, ed., Air Force Geophysics Laboratory, National Technical Information Service, ADA 167000, Chapter 5.

Butler, D. M., and Papadopoulos, K. 1984. *Solar Terrestrial Physics: Present and Future*. NASA Reference Publication 1120.

Chen, F. F. 1974. *Introduction to Plasma Physics*, Plenum Press, N.Y.

Cowley, S. W. H. 1980. The Problem of Defining a Substorm. *Nature*, vol. 286, p. 332.

Cowley, S. W. H. 1982. Substorms and the Growth Phase Problem. *Nature*, vol. 295, p. 365.

Donnelly, R. F., ed. 1979. *Solar-Terrestrial Predictions Proceedings*. U.S. Department of Commerce, Boulder, Col.

Falthammar, C. G. 1973. The Solar Wind, in *Cosmical Geophysics*, Egeland et al., eds. Universitetsforlaget, Oslo, Norway, Chapter 7.

Falthammar, C. G. 1973. Motion of Charged Particles in the Magnetosphere, in *Cosmical Geophysics*, Egeland et al., eds. Universitetsforlaget, Oslo, Norway, Chapter 9.

Greenstandt, E., Formisano, V., Goodrich, C., Gosling, J. T., Lee, M., Leroy, M., Mellott, M., Quest, K., Robson, A. E., Rodriguez, P., Scudder, J., Slavin, J., Thomsen, M., Winske, D., and Wu, C. S. 1984. Collisionless Shock Waves in the Solar Terrestrial Environment, in *Solar Terrestrial Physics: Present and Future*, D. M. Butler and K. Papadopoulos, eds., NASA Pub. 1120, Chapter 10.

Haerendel, Gerhard. 1980. The Distant Magnetosphere: Reconnection in the Boundary Layers, Cusps and Tail Lobes, in *Exploration of the Polar*

Upper Atmosphere, C. S. Deehr and J. A. Holtet, eds. D. Reidel Pubs., Co., p. 219.

Hardy, D. A., Burke, W. J., Gussehoven, M. S., Heineman, N., and Holeman, E. 1981. DMSP/F2 Electron Observations of Equatorward Auroral Boundaries and their Relationship with Solar Wind Velocity and North-South Component of the Interplanetary Magnetic Field. *J. Geophys. Res.*, vol. 86, p. 9961

Hargreaves, J. K. 1979. *The Upper Atmosphere and Solar-Terrestrial Relations*. Van Nostrand Reinhold Co., New York.

Heiles, R. A. 1982. The Polar Ionosphere. *Rev. Geophys. and Space Phys.*, vol. 20, p. 567.

Hill, T. W. 1982. Highlights of Theoretical Progress Related to the International Magnetospheric Conferences. *Rev. Geophys. and Space Phys.*, vol. 20, p. 654.

Hill, Thomas W. and Wolf, Richard A. 1977. Solar-Wind Interactions, in *The Upper Atmosphere and Magnetosphere*. National Academy of Sciences, Washington, D.C., p. 25.

Hilmer, R. V., 1989. *A Magnetospheric Magnetic Field Model with Flexible Internal Current Systems*, Ph.D. thesis, Rice University, Houston, Texas.

Horwitz, J. L. 1987. Core Plasma in the Magnetosphere, *Rev. of Geophys.*, vol. 25, no. 3, p. 579.

Huang, C. V. 1987. Quadrennial Review of the Magnetotail, *Rev. of Geophys.*, vol. 25, no. 3, p. 529.

Hundhausen, A. J. 1972. *Coronal Expansion and Solar Wind*. Springer-Verlag, New York.

Iijima, T., and Potemera, T. A. 1978. Large-Scale Characteristics of Field-Aligned Currents Associated with Substorms, *J. Geophys. Res.*, vol. 83, p. 599.

Kamide, Y., and Akasofu, S-I. 1976. The Auroral Electrojet and Field-Aligned Current. *Planet Space Sci.*, vol. 24, p. 203.

Kamide, Y., Yasuhara, F., and Akasofu, S-I. 1976. A Model Current System for the Magnetospheric Substorm. *Planet Space Sci.*, vol. 24, p. 215.

Kan, J. R. 1990. Developing a Global Model of Magnetospheric Substorms, *EOS*, vol. 71, no. 38, p. 1083.

Krall, N. K., and Trivelpiece, A. W. 1973. *Principles of Plasma Physics*, McGraw-Hill Book Co., N.Y.

National Research Council. 1981. *Solar-Terrestrial Research for the 1980s*. National Academy Press.

Nishida, A. 1978. *Geomagnetic Diagnosis of the Magnetosphere*. Springer-Verlag, New York.

Paschmann, G., and Russell, C. T. 1984. Reconnection of Magnetic Fields, in *Solar Terrestrial Physics: Present and Future*, D. M. Butler and K. Papadopoulos, eds., NASA Pub. 1120, Chapter 1.

Piddington, J. H. 1981. *Cosmic Electrodynamics* (2nd Ed). Robert E. Krieger Pub. Co., Malabar, Fla.

Rich, F. J., Gussenhoven, M. S., Hardy, D. A., and Holeman, E. 1991. Average Height-Integrated Joule Heating Rates and Magnetic Deflection Vectors Due to Field-Aligned Currents During Sunspot Minimum, *J. Atmos. and Terres. Phys.*, vol. 53, no. 3/4, p. 293.

Sagdeev, R. Z., and Kennel, C. F. 1991. Collisionless Shock Waves, *Scientific American*, April 1991, p. 106.

Schulz, M. 1976. *Plasma Boundaries in Space*. Air Force Systems Command Report SAMSO-TR-76-220.

Smith, R. E., and West, G. S. (1983). *Space and Planetary Environment Criteria Guidelines for Use in Space Vehicle Development, 1982 Revision (Volumes 1 and 2)*. NASA Technical Memorandum 82478

Space Science Board. 1978. *Space Plasma Physics: The Study of Solar-System Plasmas*. National Academy of Sciences, Washington, D.C.

Stern, A. P., and Ness, N. F. 1981. Planetary Magnetospheres. *NASA Technical Memorandum 83841*, Goddard Space Flight Center, Greenbelt, Md.

Thomas, Barry T., and Smith, E. J. 1981. The Structure and Dynamics of the Heliospheric Current Sheet. *J. Geophys. Res.* preprint.

Townsend, R. E. 1982. *Source Book of the Solar-Geophysical Environment (2nd Ed.)*. Air Force Global Weather Central, Offutt AFB, Neb.

Vasyluinas, V. M. 1974. Magnetospheric Cleft Symposium. *Trans. Amer. Geo. Union*, vol. 55, p. 60.

Vickery, J. F., Vondrak, R. R., and Mathews, S. J. 1982. Energy Deposition by Precipitating Particles and Joule Dissipation in the Auroral Ionosphere. *J. Geophys. Res.*, vol. 87, p. 5184.

Wilcox, J. M., and Ness, N. F. 1965. Quasi-Stationary Corotating Structure in the Interplanetary Medium. *J. Geophys. Res.*, vol. 70, 5793.

Wolf, R. A., Harel, M., Spiro, R. W., Voight, G.-H., Reiff, P. H., and Chen C.-K. 1982. Computer Simulation of Inner Magnetospheric Dynamics. *J. Geophys. Res.*, vol. 87, p. 5949.

Yeh, H.-C., and Hill, T. W. 1981. Mechanism of Parallel Electric Fields Inferred from Observations. *J. Geophys Res.*, vol. 86, p. 6706.

Zwickl, R. D., Bargatze, L. F., Baker, D. N., Clauer, C. R., and McPherron, R. L. 1985. *AGU Monograph on Magnetotail Physics*, American Geophysical Union, Washington, D.C.

5.13 Problems

5-1.   Describe the principal characteristics of the Earth's magnetosphere. Include information about scale size, particle density, particle energy, and current densities (if applicable).

5-2.   Differentiate between a tangential discontinuity and a contact discontinuity. What are the jump conditions for each type of discontinuity? Which discontinuity best describes the Earth's bow shock and why? Is a fast shock a reasonable explanation for substorm activity?

5-3.   The bow shock shape is determined by a balance of the external solar wind pressure (dynamic, magnetic, and gas) versus the internal magnetosheath pressure. Show that the pressure balance equation is simply approximated by

$$\frac{B^2}{2\mu_o} = nm_H v^2 \cos^2 \psi$$

where n, m_H, and v are solar wind parameters (number density, mass, and velocity) and B is the magnetosheath field strength. The angle ψ is between the velocity vector and the normal to the bow shock surface.

5-4.   Show that earthward moving tail particles (both ions and electrons) drift in such a way to produce a westward (dusk to dawn) ring current. Does this ring current add or subtract from the surface geomagnetic field measurements? What is the magnitude (current density) of the ring current?

5-5.   What is the direction of the magnetic field produced by the neutral sheet current? Similarly, what is the magnetic field direction produced by the tail current? Sketch the magnetic field configuration resulting from the combination of the neutral sheet current, tail current, and a simple dipolar main field.

5-6. Describe the principal features of the plasmasphere. What is the physical process which produces the plasmapause?

5-7. Describe the magnetospheric convection system. Does the Earth's rotation affect the final structure of the convection pattern? Does the plasmasphere participate in the magnetospheric convection? The text states that about 0.5% of the solar wind particles are needed to power the magnetosphere. Justify this statement.

5-8. What are the differences between the open and closed magnetospheric models? Does the interplanetary field orientation make any difference? Why?

5-9. Describe the time sequence of a magnetospheric substorm. Is there a difference between a magnetospheric storm and a substorm? Explain.

5-10. a. Differentiate between the growth phase model and the directly driven substorm model.

   b. The parameter ϵ is given by

   $$\epsilon = vB^2 \sin^4\left(\frac{\theta}{2}\right) L^2 \text{ (erg/sec)}$$

   where v = solar wind speed, B = solar wind magnetic field magnitude, θ = polar angle of the solar wind magnetic field vector projected on the dawn-dusk plane ($\theta = \tan^{-1}(B_y/|B_z|)$), and L = constant (~7 Earth radii). In Figure 5.20 we compare ϵ to both the AE and Dst indexes. At what approximate value of ϵ do geomagnetic storms begin? Does ϵ track very well with the storm life cycle?

5-11. Figure 5.21 contains magnetogram traces from various worldwide observation sites for a two day period starting 27 November 1959. Use these observations to answer the following questions.

   a. What physical parameters are plotted on the coordinate abscissa (i.e., what are Z, D, and H)? Which of these parameters is the best indicator for describing the start (and time development) of a magnetic storm?

   b. For the time interval 27 November to 28 November 1959, were there one or more magnetic storms? Was there a sudden commencement? If so, when?

   c. At what time did the ring current intensify? Justify your answer.

   d. What happened at College, Alaska, at 1000 UT? Was the phenomena limited to the College area alone or was there some limited global behavior?

5-12. a. Derive equation 5.29 which approximates the electric field strength in the co-rotating plasmasphere.

   b. Show the plasmapause lies approximately along the field line which maps down to about 60° magnetic latitude on the surface of the Earth. Assume the Earth's magnetic field can be approximated by a simple dipole and that the plasmapause occurs at an electric field strength of about 0.9 millivolt/meter.

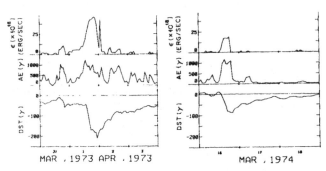

Figure 5.20   See problem 5-10 *(after Akasofu, 1982).*

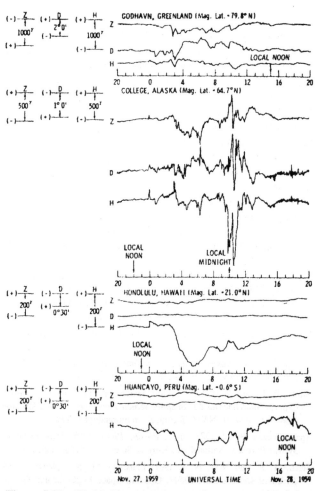

Figure 5.21   Worldwide magnetograms for a geomagnetic storm starting November 1959. See problem 5-11 *(after Akasofu, 1982).*

Chapter 6

Neutral Atmosphere

6.0 Introduction

The electrically charged portion of the atmosphere, known as the ionosphere, begins about 50 km above the Earth's surface. Above this altitude ions and electrons are present in sufficient quantities to affect the propagation of radio waves. The ionosphere is born of the interaction between solar radiation and the Earth's atmosphere and magnetic field. An understanding of the ionosphere first requires a basic knowledge of the Earth's atmospheric structure.

The atmosphere is in continual motion in response to differential solar heating. Warmer air rises, displaced by descending cool air. This overturning occurs in both vertical and horizontal planes, and is evidence that sunlight is not equally intense at all locations. Mixing, differential heating, and the Earth's rotation produce a limited stratification of the atmosphere, and it is these layers to which we first turn our attention.

6.1 Temperature Regimes

Weather, as the term is generally applied, usually occurs in the lowest 10 kilometers of the Earth's atmosphere which is known as the troposphere. This layer is characterized by a fairly steady decline in temperature (approximately 6.5 K/km) with altitude. Variable concentrations of water in the atmosphere can cause the actual variation of temperature with height to be quite different from this average.

It was believed, before the beginning of the 20th century, that the temperature continued to decline up to about 50 kilometers where the atmosphere merged into interplanetary space. However, balloon-borne measurements revealed a nearly isothermal region (about 220 K) beginning near 12 kilometers at mid-latitudes (see Figure 6.1). The level at which the temperature profile becomes isothermal is now called the tropopause. Throughout this text, transition regions will be called "pauses."

The stratosphere lies above the tropopause. In this region, the temperature is initially constant (for the first few kilometers) and then increases with height to the top of the layer. The top of the stratosphere is named the stratopause and occurs at about 45 kilometers.

Above the stratosphere is the mesosphere which is characterized by decreasing temperature with height. The top of this layer, known as the mesopause, occurs between 80 and 85 kilometers altitude. The mesopause is the coldest level of the entire atmosphere, with a temperature near 180 K.

The layer above the mesopause is the thermosphere. At this high altitude (and low density), it is important to remember that we are speaking of kinetic temperature, not sensible temperature. The thermospheric base (near 90 kilometers) marks the onset of a temperature inversion and divides the atmosphere into chemical regimes. The top of the thermosphere, the thermopause, marks a return to an isothermal temperature field. These results are summarized in Figure 6.1.

6.2 Chemical Composition Regimes

The temperature gradient establishes three distinct portions of the atmosphere—the homosphere, heterosphere, and exosphere (see Figures 6.1 and 6.2). They encompass the temperature structure outlined above.

The homosphere makes up the lower 100 km of the atmosphere. It includes the troposphere, stratosphere, and mesosphere. Vertical mixing of the atmosphere occurs in the lower atmospheric layers and keeps the relative concentrations of gases nearly constant. The homosphere is composed of (approximately) 78% molecular nitrogen, 21% molecular oxygen, and 1% argon, with variable concentrations of such gases as carbon dioxide and water vapor. This sameness in chemical composition results in the region's name.

Below the stratopause (about 30 kilometers), there is a small but significant amount of ozone, because (1) above the stratopause the concentration of molecular oxygen is insufficient to produce ozone and (2) below the ozone layer there is insufficient atomic oxygen to produce a significant amount of ozone. Thus, there is only a small region near the stratopause where there is an appreciable concentration of ozone. This layer is sometimes called the ozonosphere. From chemistry alone, one would expect ozone to be most abundant in the tropics where the photodissociation of O_2 is most efficient. However, the seasonal variation of ozone shows that the maximum abundance of ozone appears at high latitudes ($\sim 60°$ to $70°$) in the springtime because of poleward transport from

TEMPERATURE & COMPOSITION SPHERES

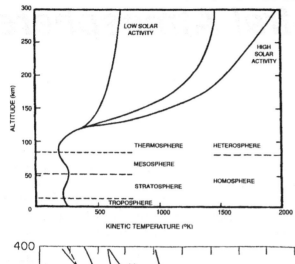

A

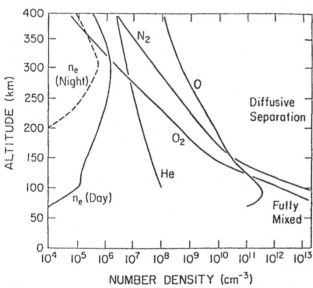

B

Figure 6.1 (A) The thermosphere is primarily heated by atomic oxygen, which absorbs EUV radiation of 1000 to 2000 A. Small contributions to the heating come from precipitating charged particles (mostly in the auroral zones) and certain chemical reactions. The energy is converted to thermospheric heat; (B) Distribution of important thermospheric constituents; above 100 km the atmospheric species distribute themselves by molecular weight. The ionosphere is imbedded in the thermosphere and electron profiles (n_e) are shown for day and night at mid-latitudes *(after Roble, 1985)*.

the tropical source regions. Transport is effective because the lifetime of individual ozone molecules can approach several weeks to a few months—long enough for the ozone to be carried along by the poleward atmospheric circulation.

Ozone absorbs virtually 100% of the incoming solar extreme ultraviolet (EUV) with a wavelength less than 2900 Å. This radiation would be lethal to many forms of life present on Earth (including unprotected man). The ozone layer shield the Earth from this radiation and heats the stratopause. Clorofluorocarbons (better known as freons under their DuPont trade name) have been implicated in the possible cata-

lytic destruction of atmospheric ozone. When the potential danger of these gases in destroying the ozone layer became known, they were banned from U.S.-made aerosol sprays even though they are still widely used as refrigerants. High levels of solar activity and high-altitude aircraft exhausts may also temporarily deplete portions of the ozone layer.

Every September the Antarctic has experienced a temporary but dramatic loss of ozone. This phenomenon, which has become popularly known as the "ozone hole" has occurred every year since it was discovered in the late 1970s. There is also an Arctic ozone hole during the northern hemisphere spring, but observations show the depletion is not nearly as severe as in the Antarctic.

The ozone abundance at any location is measured in terms of the height integrated ozone density. The resulting ozone column density is often expressed as the depth that the ozone alone would occupy at standard temperature and pressure; this depth is often measured in thousandths of a centimeter, or Dobson units, DU. Typically, ozone column abundances are about 300 DU which corresponds to a pure ozone column depth of 0.3 centimeter. On average, the highest ozone abundances approach about 460 DU at northern hemisphere sub-arctic latitudes during late springtime.

In October 1991, satellite measurements showed Antarctic abundances reached 110 DU exceeding the previous record lows of 125 DU measured in 1987, 1989, and 1990. Current theories attribute the Antarctic ozone hole to two processes. First, the deep low temperatures over Antarctica form a strong circumpolar vortex which blocks the transport of equatorial rich ozone into the polar cap region. Second, the very cold antarctic temperatures are responsible for the formation of polar stratospheric clouds which serve as the site for heterogeneous chemical reactions that liberate chlorine and bromine pollutants which then actively attack ozone molecules. Heterogeneous chemistry refers to those reactions which take place on the surfaces of liquid and solid particles in the atmosphere. When the equator-to-pole temperature contrast weakens, the southern hemisphere circumpolar vortex is disrupted allowing the influx of ozone rich equatorial air.

Northern hemispheric orographic effects allow tropospheric planetary waves to actively interact and weaken the Arctic polar vortex. As a result, warmer ozone-rich equatorial air can more readily interact with the Arctic stratosphere. In addition, the higher temperatures over the North Pole decrease the chances of stratospheric clouds from forming, and therefore, reduces the likelihood of ozone destroying chemistry from occurring.

There have been a number of studies devoted to quantifying the affect of solar EUV radiation variations associated with the solar cycle. Researchers report that ozone changes, on average, range from 1 to 4 percent from solar minimum to solar maximum.

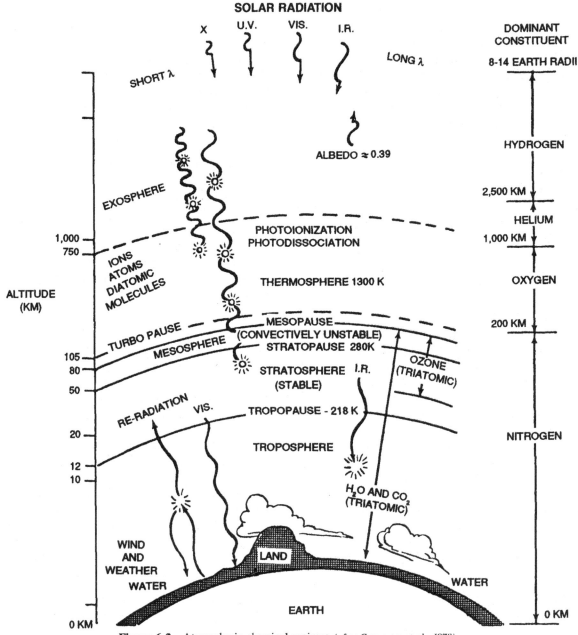

Figure 6.2 Atmospheric chemical regimes *(after Carpenter et al., 1978)*.

Above the mesopause the temperature increases steadily with height toward a value dependent on the level of solar activity (see Figure 6.1). This level is also known as the turbopause, because convective mixing ceases above this level. The absolutely stable lapse rate of this region eliminates most vertical motion, or convection. Moreover, since total atmospheric density is decreasing with height, fewer collisions occur than in the lower layers. Both of these processes minimize atmospheric mixing and allow diffusion to become important. For example, if you take a container of water and oil and mix it you get a "solution" of oil and water which will soon separate, with the heavier water molecules settling to the bottom and the lighter oil molecules rising to the top. A similar situation occurs in the atmosphere above the turbopause. Mixing stops, and the various atmospheric constituents separate out to reach hydrostatic equilibrium separately. The particles diffuse through the region, with heavier particles dominating low in the atmosphere and successively lighter constituents becoming important higher in the atmosphere (diffusive equilibrium). This layered region extending upward to 800 km is called the heterosphere (hetero is Greek for different).

The major constituents of the heterosphere are molecular nitrogen (N_2), relative weight per particle of 28; molecular oxygen (O_2), 32; atomic oxygen (O), 16; argon (Ar), 39; helium (He), 4; and atomic hydrogen (H), 1. Ordered by decreasing weight, they are Ar, O_2, N_2, and O, He, and H. In diffusive equilibrium, we would expect the heavier gases (Ar, O_2, and N_2, the dominant gases of the troposphere) to be relatively unimportant at higher levels, where the lighter gases, especially atomic hydrogen, tend to gather.

Starting within the center of the heterosphere, the thermosphere is primarily heated by atomic oxygen, which absorbs EUV radiation with wavelength of 1000 to 2000 angstroms. Appreciable contributions to the heating come from precipitating charged particles (mostly in the auroral zones) and certain chemical reactions.

Above the heterosphere, the exosphere extends upward to nearly 1000 km. The exosphere is a region of continual loss of atmospheric particles due to long mean free paths and high kinetic energies (temperatures). The very low particle density at high altitudes means the average distance between particles is large. At 800 kilometers, for example, the mean free path of atmospheric particles is approximately 160 kilometers. A neutral particle in the Earth's atmosphere is restrained from escaping by only two things: gravity and collisions. In the upper thermosphere, the high kinetic temperature means that a large portion of the particles are traveling at speeds greater than escape velocity. So long as they remain electrically neutral these particles can easily escape unless they collide with another particle, give up energy, and slow to below escape velocity. The large mean free path of the upper thermosphere makes collision probability very low, so many of these particles escape the Earth. Those traveling too slow to escape may later absorb incoming solar EUV radiation and reach a kinetic temperature high enough to escape.

Lighter particles, such as hydrogen and helium, are traveling at a higher speed than heavier particles (N_2, O_2, Ar) at the same kinetic temperature, and so escape more readily from lower altitudes than do heavier particles. Because of this, there is no fixed lower boundary of the exosphere, and it is generally thought to be somewhere between 500 and 1000 kilometers. Hydrogen and helium dominate at this level, and they are steadily lost from the exosphere. Variations of exospheric temperature occur due to such factors as solar activity, time of day, and latitude.

6.3 Density Variations

Density variations may have a significant impact on radio wave propagation because increasing the particle density between 60 and 100 km (or higher) can alter the Earth's ionosphere by providing more material for ionization. The atmospheric density at high altitudes changes in response to many factors including local time, latitude, altitude, and

level of solar and geometric activity. For example, since the amount of solar radiation (insolation) received at a point on the Earth's surface is a function of local time, it determines to a large extent the atmospheric temperature and therefore the air density. Insolation also falls off rapidly with distance from the subsolar point (the point on the Earth's surface directly between the center of the Earth and the Sun), and thus produces a latitudinal effect on the observed atmospheric temperature for different seasons.

Averaging the density over local time and latitude to remove Sun angle effects reveals that the average worldwide density varies in a cyclic fashion with a semiannual period. The highest average density occurs in October with a secondary peak in April. The lowest average density occurs in July with a secondary minimum in January. This effect is actually the result of two features: the variable Earth-Sun separation and the higher solar zenith angle (and longer day) in the summer hemisphere.

The Sun's EUV output varies in a pattern similar to sunspot number (SSN), and this variability translates into a variation of energy available to the thermosphere. The resulting variation of exospheric temperature, in turn, produces a solar cycle variation of atmospheric density. Little EUV radiation reaches the ground, and direct EUV flux observations have been made only rarely. However, we can infer the value based on 2800 MHz solar radio flux measurements because EUV and 2800 MHz fluxes show a fairly good correlation. The 2800 MHz flux is better known as the 10.7 centimeter flux (or F10). Although the correlation is not perfect (and varies from one sunspot cycle to the next), the patterns are similar enough to be useful. During solar minimum conditions (Cycle 21), a sunspot number of approximately 15 corresponded to an F10.7 of about 70 flux units (Note: a flux unit is $10^{-22} Wm^{-2}Hz^{-2}$). A F10.7 flux of 200 corresponds to a sunspot number of about 150 to 160.

Daily values of 10.7 cm radio flux are observed near Ottawa, Canada, at 1400Z, 1700Z, and 2000Z. There is some atmospheric attenuation of the received flux, dependent on Sun angle, with the minimum at local noon. The 17Z (noon) flux value is the world standard and is archived in a manner similar to SSN. These values are used as "measurements" of the solar EUV radiation which heats the atmosphere. Two 10.7 cm flux values are used in most atmospheric density models, the daily value and a longer term mean, to act as a surrogate parameter for the EUV flux.

When a geomagnetic storm occurs, large numbers of charged particles are dumped from the magnetosphere into the high latitude atmosphere. These particles ionize and heat the high latitude atmosphere by collisions, with the heating first observed several hours (1 to 10) after the geomagnetic disturbance begins. The effects of geomagnetic heating extend from at least 300 km to over 1000 km and may persist for 8 to 12 hours following the end of the magnetic disturbance.

6.4 Stratospheric Dynamics

At low stratospheric altitudes (below 21 km) there is considerable energy and mass exchange between the stratosphere and troposphere within a given hemisphere. Circulation models show that during the winter, the low-latitude tropospheric Hadley cell transfers tropospheric air upward and poleward at a rate of about 8×10^9 kg/sec between the tropopause and the 25 km level. The Hadley cell is the intense mass circulation between the equator and 30°N with warm air rising at the equator and colder air descending in the north. Altogether, the Hadley cell transfers 35 to 40% of the mass equivalent of the stratosphere through the tropopause per year. The mass transfer near the equator occurs in tropical updrafts which penetrate the tropopause. In addition, water vapor is carried into the stratosphere in the towering cumulus clouds of subtropical and tropical thunderstorms. Residence time for stratospheric particles ranges from a few days to more than a year.

Above 35 km, stratospheric flow is dominated by zonal flow and the exchange of mass between hemispheres (north and south) becomes important at these altitudes. In response to the solar heating in the summer hemisphere and radiational cooling in the winter hemisphere, a flow is set up in the stratosphere which carries mass across the equator into the cold winter hemisphere. The total mass transferred in 6 months is about 10^{17} kilograms, or approximately 5 to 10% of the total mass of the stratosphere. The flow into the winter hemisphere can disrupt and abruptly change the planetary flow pattern allowing stronger downward motion. The sinking high-latitude air heats up, and the polar stratosphere's temperature may increase by up to 100° C. The phenomena is called "sudden stratospheric warming" because the temperature increase occurs over a period of weeks. Historically, such warmings were thought to occur aperiodically and thus the sudden appearance was usually a surprise. However, with today's modern satellite instrumentation we have found that the winter heating is indeed a periodic seasonal effect which is evident especially in the mesosphere. Only very intense periods of warming reach down into the stratosphere, and thus, the periodic nature of the warming phenomena is lost to observers whose instrumentation limits measurements to only the lower stratosphere. Stratospheric warmings are most intense in the northern hemisphere, and observations show the weaker southern hemisphere warmings occur later in the wintertime than in the north.

In the intermediate altitude range, 21–35 km, there is little exchange between the summer and winter hemispheres. About the equator, there is an unusual perennial reversal of the low altitude stratospheric winds which is called the quasi-biennial oscillation. The wind reversal first appears at high elevations and gradually propagates downward. Unfortunately, the mechanisms of these phenomena are not well understood.

6.5 Mesospheric Dynamics

The mesosphere is characterized by the coldest atmospheric temperature ($-92°$ C) at the mesopause. The average decrease of temperature with height is about 3° per km, about half the rate of the troposphere. This cooling is due to "vibrational relaxation" of carbon dioxide molecules. That is, the mesosphere cools by the spontaneous emission of photons from an excited vibrational state of CO_2. The excited state is populated by collisions because above 50 km altitude, radiative absorption is negligible compared with collisional excitation. Since the spontaneous emission occurs faster than the excitation, the mesospheric gas continues to cool until downward conductive heating from the overlying thermosphere offsets the radiant losses.

Above 45 km, the most striking feature of the global circulation is a reversal of the zonal wind direction from easterly in the summer to westerly in the winter. Again, zonal refers to the average west-to-east or east-to-west flow which predominates the upper atmospheric circulation. The easterlies reach their peak in midsummer and the westerlies are at maximum after the winter solstice. The spring and fall transition periods occur about a month after their respective equinoxes.

6.6 Thermospheric Dynamics

Above 80 km solar ultraviolet radiation is particularly efficient in the photodissociation of molecular oxygen into atomic oxygen. Atomic oxygen recombines very slowly above 90 km, so the upper thermosphere is rich in atomic oxygen. Of the total solar energy absorbed in the thermosphere, approximately 60% heats the ambient neutrals, 20% is radiated away as ultraviolet airglow, and the remainder is available for atomic oxygen chemistry.

The thermospheric temperature is controlled by competition of heat sources balanced by heat losses. The primary heat source is the absorption of UV/EUV radiation by the neutral constituents making up the thermosphere; the total EUV energy input from 150 to 400 km is approximately 8×10^{10} watts, and the UV energy input from 90 to 150 km altitude is approximately 10^{12} watts. In addition, important energy sources at high latitudes include magnetospherically driven Joule heating and particle precipitation. The principal loss processes include downward heat conduction, and photoemissions produced as molecules (and atoms) relax from their excited vibrational and radiational states. Figure 6.3 shows the heating and cooling rates for various sources and sinks as a function of altitude. One minor source (~5%), not shown on Figure 6.3, is due to photoionization. Newly created primary photoelectrons can lose energy either by coulomb collisions with other charged particles or by inelastic collisions with neutral atoms or molecules. However, these processes only net minor heating because most of the transferred energy

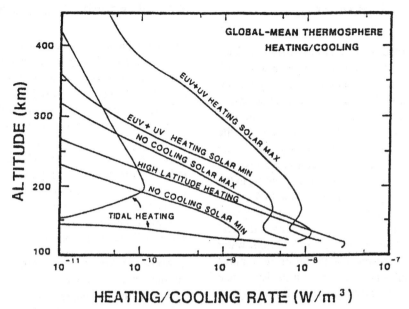

Figure 6.3 Average global thermospheric heating and cooling rates as a function of altitude. High latitude energy source is due to both magnetospherically driven joule heating and particle precipitation. The main cooling mechanism (not shown) is the slow process of downward heat conduction *(after Killeen, 1987)*.

is lost to airglow emissions created as energized molecules and atoms (neutral and charged) relax to lower energy states.

Figure 6.3 shows that the heating and cooling rates increase with decreasing altitude. In large part, the heating increase is due to the efficient absorption (~33%) of the 1300 to 1750 Å Schumann-Runge solar radiation continuum. In this spectral band, solar photons dissociate molecular oxygen O_2, producing metastable atomic oxygen in an excited energy state. Since the energy state is metastable, the oxygen atoms have sufficient time to redistribute their excess energy via collision with neighboring neutral atoms and molecules. However, the increasing collision rate with decreasing altitude is a two edged sword. As neutral molecules collide with metastable oxygen, the transferred energy often produces excited molecular vibrational states which quickly relax via photoemissions.

Changes of the cooling and heating rates with altitude are usually due to compositional and density variations which directly affect the energy transfer efficiencies. Take, for example, the difference in nitric oxide (NO) cooling rates which occur between solar maximum and solar minimum. Studies show that during solar minimum, the solar UV/EUV heating is balanced by downward heat conduction. Molecular cooling by NO is less important because, at solar minimum, thermospheric NO concentrations are relatively small. However, at solar maximum, the 2 to 3 times increase in EUV radiation produces higher thermospheric temperatures which increases NO concentrations as the lower thermosphere swells, and in turn, makes NO molecular cooling an important loss term.

The absorption of solar electromagnetic radiation is normally the dominant forcing mechanism in the thermosphere. For example, the thermosphere has a diurnal temperature variation of nearly 30%, with the maximum value near the subsolar point and the minimum value near the antisolar point. Furthermore, the diurnal temperature variations depend on latitude and season which are consistent with the hypothesis that solar radiation (especially UV/EUV) is the principal factor in determining the thermospheric temperature. Figure 6.4 shows the variation of mid-latitude thermospheric mass density during spring equinox at solar maximum and active geomagnetic conditions; Figure 6.5 shows the variation of thermospheric temperature for the same conditions. Similarly, Figures 6.6 and 6.7 show how mass density and temperature vary as a function of month for the same geophysical conditions. The results shown in these figures are from an empirical thermospheric model (Hedin, 1983) developed from mass spectrometer and incoherent scatter radar (MSIS) data.

There are also longer term variations due to the fact that the amount of solar UV/EUV depends on the 27 day rotation period of the Sun and the 11 year solar cycle. The principal sources of solar UV/EUV are active regions which rotate on and off the visible solar disk. Furthermore, the number of active regions varies with the 11 year solar cycle (see Chapter 2).

The low-latitude thermosphere is primarily controlled by solar radiation, atmospheric tides, and magnetospherically driven changes in the equatorial ionospheric electric fields. In fact, the ion drift measurements on the geomagnetic

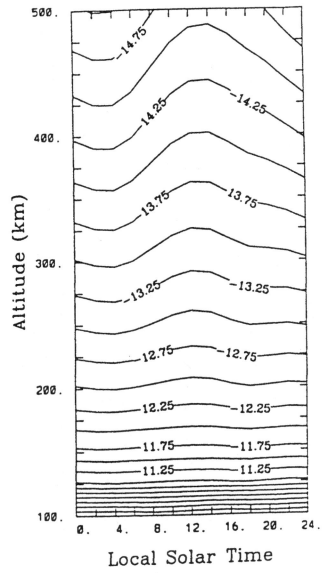

Figure 6.4 Comparison of mid-latitude thermospheric mass density variation with altitude versus local time generated by the MSIS empirical thermospheric model by Hedin, 1983. The model was run for solar maximum (F10.7 = 180) during active geomagnetic conditions (A_p = 13 or K_p = 3) on Julian day 100. Mass density units are 10^{-14} g/cm^3 *(after Killeen and Raskin, 1991).*

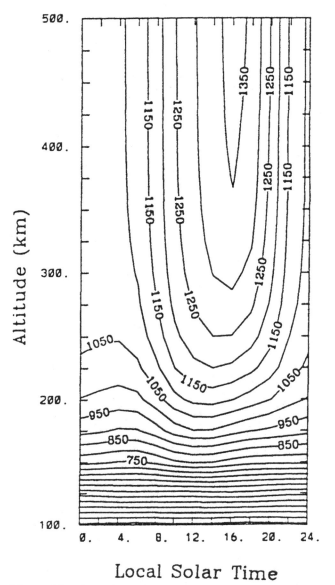

Figure 6.5 Comparison of mid-latitude thermospheric temperature variation with altitude versus local time generated by the MSIS model. The model was run for solar maximum during active geomagnetic conditions on Julian day 100. Units are in degrees kelvin *(after Killeen and Raskin, 1991).*

equator show that the normal east-west equatorial electric field may be disturbed for 16 to 24 hours after geomagnetic storm onset. The disrupted electric field is maintained by the motion of the "spun-up" neutral atmosphere which, like a fly-wheel, takes time to relax to prestorm conditions.

The mid-latitude thermosphere above 200 km is controlled primarily by solar UV/EUV radiation, but it can be influenced by high-latitude phenomena during intense geomagnetic storms. The high-latitude convection electric field is pushed equatorward during strong geomagnetic storms disrupting the normal mid-latitude flow pattern. In addition, areas of significant joule heating form thermospheric bulges

which propagate equatorward as atmospheric gravity (coherent density) waves. For example, observations show that strong geomagnetic disturbances produce a post-midnight equatorward density surge and strong westward winds consistent with the expected ionospheric circulation.

As expected, the high-latitude thermosphere is more dynamic and variable than either the low- or mid-latitude regions. Geomagnetic storms act as a significant heat source in two ways. First, the precipitating charged particles (and secondary electrons) can energize neutrals via direct inelastic collisions, and the energized neutrals in metastable states can, in turn, then collisionally redistribute their excess en-

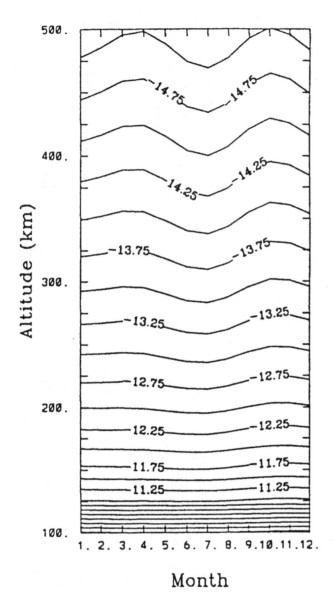

Month

Figure 6.6 Comparison of mid-latitude mass density variation with altitude versus month of year for the MSIS model. Computation for solar maximum during active geomagnetic conditions at 0400 universal time (UT). Mass density units are 10^{-14} g/cm³ *(after Killeen and Raskin, 1991).*

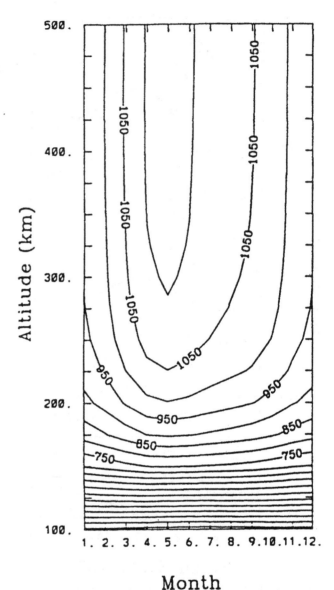

Month

Figure 6.7 Comparison of mid-latitude temperature variations with altitude versus month of year for the MSIS model. Computation for solar maximum during active geomagnetic conditions at 0400 universal time (UT) *(after Killeen and Raskin, 1991).*

ergy. Second, changes in the magnetosphere induce changes in the ionospheric polar electric field which then enhances and redistributes the polar vortex (see Chapter 5) transferring energy (joule heat) to the thermosphere via ion drag. The resulting polar vortex formed in the thermosphere is superimposed on the strictly solar UV/EUV driven antisunward transpolar flow. Satellite observations show the dominance of the anticyclonic neutral vortex associated with the dusk cell of the ion convection pattern. During disturbed geomagnetic periods, polar cap winds can reach 1 km/sec approaching the sonic velocity. There is also observational evidence that the vertical component of these winds can exceed 100 m/sec. Figure 6.8 shows the 350 km altitude ion drift pattern

deduced from satellite measurements. As mentioned above, the neutral particles are collisionally coupled with the ions such that the neutral winds parallel the ion convection. In both the morning and evening sectors, the transition from corotation to the large sunward convection velocities begins near the edge of the diffuse aurora. Section 5.4 describes the structure and variability of the electric field pattern driving the ion convection pattern shown in Figure 6.8.

Research studies show a reasonable linear correlation exists between AE and the strength of the high-latitude joule heating; the proportionality constant is 2.0×10^{8} watts per unit of AE. A useful, but less accurate, proportionality

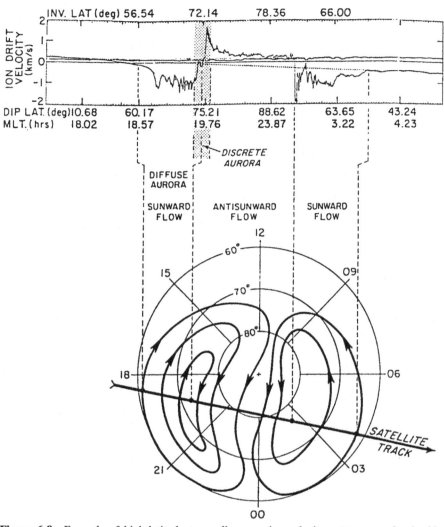

Figure 6.8 Example of high-latitude two cell convection velocity pattern associated with southward IMF. Pattern drawn from satellite measurements taken at 350 km altitude. The panel at the top of the figure shows the satellite measurements and the relative location of the diffuse and discrete aurora *(after Ossakow et al., 1984).*

constant was also found for the heat contribution from precipitating geomagnetic storm particles; the linear proportionality constant is 0.6×10^8 watts per unit of AE. Observations also show that joule heating occurs roughly in an oval pattern consisting of three distinct heating regions: the dayside cleft, the region of sunward ion convection at dawn, and the midnight sectors. There is also evidence that joule heating in the summer hemisphere can be significantly greater ($\sim$50%) than that in the winter hemisphere because of enhanced summer ionospheric conductivities.

Energy can also be transferred to the thermosphere from below by upward propagating tides and gravity waves generated at lower altitudes (see Figure 6.3). These waves have large amplitudes because they tend to grow exponentially as they propagate upward into increasingly rarefied regions. Waves can have periods ranging from minutes to one day; periods of 12 to 24 hours are produced by the daily variation

of solar UV absorption in the ozone layer. Such waves are called thermal waves and can produce winds of around 100 m/sec as they propagate upward into the thermosphere. Overall, tidal heating contributes about 7% of the total thermospheric energy input.

6.7 Summary

The Earth's neutral atmosphere is vertically differentiated by composition, density, and temperature. Atmospheric dynamics are controlled by pressure gradients resulting from differential solar heating, and by gravity. The high-altitude atmosphere becomes coupled with the ionosphere due to ion-neutral collisions. Therefore, the electrodynamics of the ionosphere helps shape the neutral atmosphere, and conversely, high-altitude neutral wind dynamics can play a significant role in determining ionospheric structure.

6.8 References

Balachandran, N. K. 1983. Acoustic and Gravity Waves in the Neutral Atmosphere and the Ionosphere, Generated by Severe Storms. *NASA Conference Publication 2259*, p. 12.

Chamberlain, J. W. 1978. *Theory of Planetary Atmospheres*. Academic Press, New York.

Craig, R. A. 1965. *The Upper Atmosphere: Meteorology and Physics*. Academic Press, New York.

Danielsen, E. F., and Louis, J. F. 1977. Transport in the Stratosphere, in *The Upper Atmosphere and Magnetosphere*. National Academy of Sciences, Washington, D.C.

Donahue, T. M. 1977. Hydrogen (Models of Exosphere), in *The Upper Atmosphere and Magnetospheres*, Chapter 4. National Academy of Sciences, Washington, D.C.

Haggard, K. V., and Grose, W. L. 1981. Energetics of a Sudden Stratospheric Warming Simulated With a Three-Dimensional, Spectral, Quasi-Geostrophic Model, NASA Technical Paper 1847.

Hedin, A. E. 1983. A Revised Thermospheric Model Based on Mass Spectrometer and Incoherent Scatter Data: MSIS-83, *J. Geophys. Res.*, vol. 88, p. 10170.

Killeen, T. L. 1987. Energetics and Dynamics of the Earth's Thermosphere, *Reviews of Geophysics*, vol. 25, no. 3, p. 433.

Killeen, T. L., and Raskin, R. 1991. Vector Spherical Harmonic Model, U.S. Air Force Space Environmental Models Conference, Colorado Springs, Col.

Kim J. S., Murty, G. S. N., Kim, J. W., and Kim, Y. 1990. Thermospheric Temperature During a High Solar Activity Period, *EOS*, vol. 71, no. 39, p. 1100.

Miller, A., and Thompson, J. C. 1975. *Elements of Meteorology (2nd Ed)*. Charles E. Merrill Publishing Company, Columbus, Ohio.

Monastersky, R. 1990. Antarctic Ozone Hole Returns With a Bang and Antarctic Ozone Bottoms at Record Low, *Science News*, vol. 138, pp. 198, 228.

Moyer, V. 1976. *Physical Meteorology*. Texas A & M University, College Station, Tex.

Ossakow, S. L., Burke, W., Carlson, H. C., Heelis, R., Keskinen, M., Maynard, N., Meng, C., Szuszczeweiz, E., and Vickery, J. 1984. High Latitude Ionospheric Structure, in *Solar Terrestrial Physics: Present and Future*, D. M. Butler and K. Papadopoulos, eds., NASA Ref. Pub. 1120, Chapter 12.

Prochaska, R. D. 1980. *Space Environmental Forecaster Course*. Air Force Global Weather Central, Neb.

Roble, R. G. 1977. The Thermosphere, in *The Upper Atmosphere and Magnetosphere*. National Academy of Sciences, Washington, D.C.

Roble, R. G. 1985. The Response of the High-Latitude Thermosphere to Auroral Processes, AIAA/Astrodynamics Conference, Vail, Col.

Roble, R. G., and Killeen, T. L. 1985. Measurements of the High-Latitude Thermosphere, AIAA/Astrodynamics Conference, Vail, Col.

Salby, M. L., and Garcia, R. R. 1990. Dynamical Perturbations to the Ozone Layer, *Physics Today*, March 1990, p. 38.

Schoeberl, M. R. 1987. Dynamics of the Middle Atmosphere, *Reviews of Geophysics*, vol. 25, no. 3, p. 501.

Townsend, R. E. 1982. *Source Book of the Solar-Geophysical Environment*, Air Force Global Weather Central, Neb.

Wuebbles, D. J. 1987. Natural and Anthropogenic Perturbations to the Stratosphere, *Reviews of Geophysics*, vol. 25, no. 3, p. 487.

6.9 Problems

6-1. Why is the ozone layer important for life on Earth? Why does the layer form in the lower stratosphere? Does the ozone concentration change with the seasons? Is the ozone hole observable throughout the entire year? Explain.

6-2. Why are thermospheric winds important for the ionosphere? Do you think thermospheric winds can affect high frequency radio communications?

6-3. Would the thermosphere feel hot to you? Explain what is meant by thermospheric temperature. What are the physical processes behind the daily variation in thermospheric temperature? How about the monthly variations in thermospheric temperature? Are the same processes responsible for the annual variations?

6-4. Do upper atmospheric dynamics above 15 km have any effect on the troposphere? The stratosphere has measurable amounts of water vapor. Where does this water vapor come from? Is it seasonal?

6-5. What is the sonic velocity in the thermosphere? Does the sonic velocity change with latitude? Longitude? Day, month, or year? Are the thermospheric winds ever supersonic? Explain.

6-6. Describe the chemical makeup of the atmosphere between 20 and 100 km.

6-7. Do mesospheric dynamics have any impact on the ionosphere? If so, why?

6-8. Describe how the neutral atmospheric density and temperature change with increasing altitude. Do the temperature and density change with time, and if so, what are the underlying processes behind such changes?

6-9. Using the AE trace in Figure 5.16, compute the approximate joule heating in the auroral zone for the geomagnetic storm in March 1973. How does this energy input compare to the solar radiation energy input during this same period of time?

Chapter 7

Ionosphere

7.0 Introduction

The radiation from the Sun contains sufficient energy at short wavelengths to cause appreciable photoionization of the Earth's atmosphere at high altitudes resulting in a partially ionized region known as the ionosphere. Within the ionosphere, the recombination of the ions and electrons proceeds slowly (due to low gas densities) so that fairly high concentrations of free electrons persist even throughout the night. In practice, the ionosphere has a lower limit of 50 to 70 km and no distinct upper limit, although 2000 km is somewhat arbitrarily set as the upper limit for most application purposes.

The existence of a conducting region in the Earth's upper atmosphere was a subject of early speculation by William Thomson (Lord Kelvin) and was invoked by Balfour Stewart in 1882 in connection with daily magnetic variations. However, it was Guglielmo Marconi's demonstration of long distance radio communication in December 1901 that stimulated widespread studies of this phenomenon. In 1902 Arthur Kennelly and Oliver Heaviside independently postulated an ionized atmospheric layer to explain the radio transmission at a time when others were investigating diffraction effects as the explanation. A decade later, William Eccles supplied the rudimentary theory of how radio waves propagate through an ionized medium. The problem was taken up again by John Larmor in 1924, and his work still gives a good elementary, first-approximation explanation of the reflection of radio waves from an ionized region. Direct evidence for the existence of the "Kennelly-Heaviside layer" (currently called the E-layer) was obtained by comparing the fading of signals received on two types of directional antennas. This work was done by Appleton and Barnett (1925) and Smith-Rose and Barfield (1927). In 1926, Breit and Tuve developed a direct method of probing the ionosphere by measuring the round-trip for reflected, ground-launched radio waves. A variation of this method is the primary method used today for probing the ionosphere.

Since the 1920s we have discovered that the ionosphere varies dramatically with geomagnetic latitude. In particular, there appear to be three distinct ionospheric regions: high-latitude, midlatitude, and low-latitude. The midlatitude ionosphere is easiest to understand and most closely follows the classical ionospheric models. The high-latitude iono-sphere is directly coupled to the magnetospheric tail by the stretched auroral magnetic field lines, and this connection has important consequences for the high-latitude ionosphere. The low-latitude ionosphere is sensitive to plasma instabilities and changes to the magnetospheric ring current. In this chapter we will first develop the classic ionospheric model and in Chapter 8 we will apply it to each geographic region to determine the model limitations, modifications (where possible), and improvements.

7.1 Vertical Electron Density Profile of the Ionosphere

The vertical structure of the ionosphere is changing continuously. It varies from day to night, with the seasons of the year, and with latitude. Furthermore, it is sensitive to enhanced periods of short-wavelength solar radiation accompanying solar activity. Still, the essential features of the ionosphere are usually identifiable, except at high latitudes during periods of unusually intense geomagnetic disturbances. The different ionospheric vertical layers are shown in Figure 7.1. In order of increasing altitude and increasing electron concentration, these layers are called D, E, F1, and F2. Figure 7.1 also shows how typical daytime and nighttime vertical electron density profiles change over the course of the sunspot cycle (profiles apply for midlatitudes only). Above the maximum electron density of the F2-region, the electron density decreases monotonically out to several Earth radii. Not only does the overall electron density decrease at night, but the F1- and D-layers disappear soon after sunset.

Distinct ionospheric regions develop because (a) the solar spectrum deposits its energy at various heights depending on the absorption characteristics of the atmosphere, (b) the physics of recombination depends on the atmospheric density (which changes with height), and (c) the composition of the atmosphere changes with height. Thus, the four main ionospheric regions can be associated with different physical processes, rather than simple height differentiation.

The electron density is approximately equal to the ion density everywhere in the ionosphere; this property is referred to as charge neutrality. The exception to this equality is in the D-region, where the electrons may combine with molecules to form negative ions (such ions are of no impor-

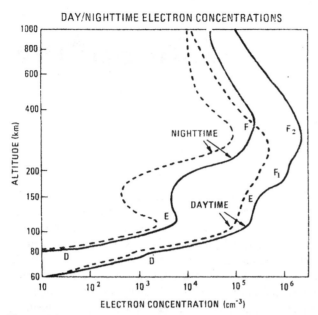

DAY/NIGHTTIME ELECTRON CONCENTRATIONS

Figure 7.1 Typical midlatitude daytime and nighttime electron density profiles for sunspot maximum (solid lines) and minimum (dashed lines).

tance elsewhere in the ionosphere). Above the D-region charge neutrality results because ions and electrons are nearly always created and eliminated in pairs. Usually, electric forces prevent the free electrons from wandering too far away from their parent ions. As we'll soon see, charge separation is possible in areas where the ion collision frequency is high, and the more mobile electrons produce ionospheric currents which help shape the observed ionospheric configuration.

It has become common practice to speak of the ionospheric layers in terms of a specific radio frequency rather than the electron number density. This radio frequency (called the critical frequency) is the highest frequency which is reflected by the layer in question for normal incidence. The maximum electron density is related to critical frequency by $N_{e_{max}} = 1.24 \times 10^4 \, f^2$ where N_e (cm^{-3}) is the maximum electron density and f is the critical radio frequency expressed in megacycles per second. In plasma physics, the critical frequency is known as the plasma frequency, but critical frequency is the term usually used in ionospheric work.

7.2. Simple Chapman Layer

The ultraviolet solar energy incident on the upper atmosphere is absorbed at various altitudes and by various gaseous constituents before reaching the lower atmosphere. The "law" that this absorption follows (theoretically) is given by

$$I_{\nu s} = I_{\nu 0} \exp\left\{-\int_0^s k_\nu \rho \, ds\right\}. \qquad (7.1)$$

where $I_{\nu s}$ is the specific intensity (units are joules/m^2-sec-sr-Hz) along a specified direction, s is usually taken to be slant distance, $I_{\nu 0}$ is the value of the specific intensity at some

point where s is taken to be zero, k_ν is the absorption coefficient, and ρ is the density of the medium. The quantity represented by the integral usually is called the slant optical thickness.

Sydney Chapman first treated the problem of ionospheric formation. Although his results were based on a number of simplifications and his treatment has been extended and generalized in various ways, he captured the essence of the problem. The idealized problem treated by Chapman involves the absorption of a beam of parallel, monochromatic radiation impinging on an atmosphere of uniform composition in which the density varies exponentially with height. First we will consider the absorption as a function of height and of the zenith angle of the incoming radiation. Second, we will take the Earth to be flat, solely because we do not want the derivation obscured by foggy geometrical complications.

We will let the flux of radiation be F and the zenith angle be θ. Then we propose that the density of the air ρ varies with height z according to:

$$\rho = \rho_0 \exp(-z/H), \qquad (7.2)$$

where H is the so-called "scale height" [$H = kT/mg$, where k is Boltzmann's constant, T is Kelvin temperature, m is the molecular weight of the medium, and g is the acceleration due to gravity], and ρ_0 is the value of ρ at some arbitrary level at which z is taken to be zero. The flux F is related to the intensity by

$$\pi F_\nu = 2\pi \int_{-1}^{1} I_\nu(\mu) \, \mu \, d\mu \qquad (7.3)$$

where $\mu = \cos\theta$ and the subscript ν denotes frequency dependence. Equation 7.3 gives the net flux (energy per unit area per second) across an area, integrated over all incident beams, coming from all directions. Equation 7.3 is commonly written in terms of the mean specific intensity

$$\pi F_\nu = \pi I_{mean} \qquad (7.4)$$

where we have assumed no outward atmospheric radiation.

In order to simplify our notation, we will let τ be the vertical optical depth where

$$d\tau_\nu = -k_\nu \rho \, dz \qquad (7.5)$$

and the subscript ν implies frequency dependence of the incident radiation which is being absorbed by the atmosphere. The slant range s can be written in terms of the vertical height z, and zenith angle θ by the equation

$$ds = \sec\theta \, dz = \frac{dz}{\mu}. \qquad (7.6)$$

In terms of equations 7.4 and 7.5, equation 7.1 becomes

$$\frac{dI_\nu}{d\tau_\nu} = \frac{I_\nu}{\mu} \qquad (7.7)$$

where we have used the differential form for later convenience. Equation 7.7 assumes that the thermal emissions and scattering are unimportant. If one wanted to include these effects then equation 7.7 becomes

$$\mu \frac{dI_\nu}{d\tau_\nu'} = I_\nu - J_\nu \qquad (7.8)$$

where J_ν is called the source function and $d\tau_\nu' = (k_\nu + \sigma_\nu)\,ds$ where σ_ν is the mass scattering coefficient. The general derivation and use of equation 7.8 is described more fully in Chamberlain (1978).

In terms of the mean intensity, equation 7.7 becomes

$$\frac{dF_\nu}{d\tau_\nu} = \frac{F_\nu}{\mu}. \qquad (7.9)$$

Integrating and using the definitions of equations 7.2, 7.5, and 7.6 we find

$$F_\nu = F_\infty \exp\left(-k_\nu \rho_o H \sec\theta \exp\left(-\frac{z}{H}\right)\right) \qquad (7.10)$$

where F_∞ is the flux at the top of the atmosphere. Notice that equation 7.10 is an exponential within an exponential. Remember that F_ν is the incident energy per unit area per unit time per unit frequency.

The rate of photoionization due to this incident energy will depend on the amount of incident energy per unit volume times the total number of atoms per unit volume available to absorb the radiation. The amount of incident energy per unit volume per unit time is simply given by $dF_\nu/(\sec\theta\,dz)$. If we use $N(z)$ to denote the atmospheric density, the photoionization production rate (photons/sec m^3) becomes

$$q_\nu = \frac{1}{E_\nu \sec\theta}\frac{dF_\nu}{dz} = N(z)\frac{F_\nu}{E_\nu}k_\nu m = N(z)\mathbf{F}_\nu K_\nu \qquad (7.11)$$

where E_ν is the energy per photon, $\mathbf{F}_\nu = F_\nu/E_\nu$, $K_\nu = k_\nu m$, and we also have used equations 7.5 and 7.9. Since the density decreases with height (equation 7.2), the number density must also decrease in the same way because we initially assumed that the atmosphere has a uniform composition. Therefore, using equations 7.2 and 7.10, we can write the production rate as

$$q_\nu = K_\nu N_o \mathbf{F}_\infty \exp\left[-\left(\frac{z}{H}\right) - k_\nu \rho_o H \sec\theta \exp\left(-\frac{z}{H}\right)\right] \qquad (7.12)$$

where N_o is the number density at the base of the atmosphere.

At high altitudes, q_ν decreases because $N(z)$ is decreasing with altitude. At low altitudes, q_ν again decreases because almost all of the incident radiation has been absorbed by the overlying atmosphere. Thus there must be some intermediate level, z_m, where q_ν is a maximum. We find that the maximum production height is given by

$$\exp\left(\frac{z_m}{H}\right) = k_\nu \rho_o H \sec\theta \qquad (7.13)$$

and the maximum production rate is

$$q\,(\text{max}) = q_m = \frac{\mathbf{F}_\infty \cos\theta}{H}e^{-1}. \qquad (7.14)$$

It is interesting to note that z_m is independent of the initial flux in the beam and q_m is independent of the absorption coefficient. With regard to zenith angle, z_m increases and q_m decreases as θ increases. It is convenient to express q_ν as a function of the distance from the level of maximum absorption (i.e., the peak in the "electron density" curve); that is, as a function of a variable z_1, where

$$z_1 = (z - z_m)/H. \qquad (7.15)$$

With this variable, equation 7.12 becomes

$$(q/q_m) = \exp\left[1 - z_1 - \exp(-z_1)\right]. \qquad (7.16)$$

If we plot this function, we will see that most of the absorption takes place within four scale heights above and two scale heights below the level of maximum absorption. The shape of the layer is independent of zenith angle, except in that q_m (and therefore q) varies as $\cos\theta$ according to equation 7.14. Figure 7.2 shows the distribution of electron density as a function of height for various zenith angles.

It is phenomenologically correct to treat the ion-electron recombination rate as proportional the square of the electron density. Therefore the rate of change of the ionospheric

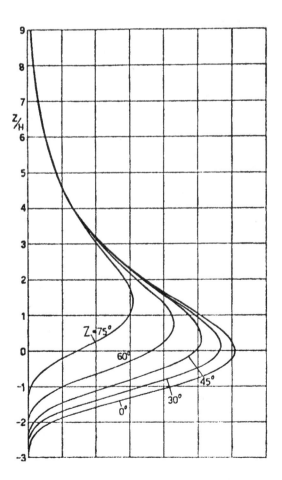

Electron Density

Figure 7.2 Distribution of electron density versus zenith angle *(after Rawer, 1952).*

electron density is simply the difference between the production and loss (recombination) rates or

$$\frac{dN_e}{dt} = q_\nu - \alpha_{eff} N_e^2 \qquad (7.17)$$

where α_{eff} is the effective recombination coefficient. Table 7.1 outlines the ionospheric production and loss processes and the characteristic values of α_{eff} (determined experimentally) for the principal ionospheric layers.

Using equation 7.12 together with equation 7.17, we find that the electron density at any height for a "Chapman type" ionosphere is given by

$$N_e(z, \theta) = \left(\frac{q_m}{\alpha_{eff}}\right)^{1/2} \exp\{\tfrac{1}{2}[1 - z_1 - \exp(-z_1)]\}. \qquad (7.18)$$

There are two important characteristics for this idealized Chapman layer. First, expanding the exponential about z_m for small distances compared with H (assuming overhead Sun, $\theta = 0$) shows that the electron distribution is parabolic around the peak of the layer. Secondly, the daily, seasonal, and latitudinal variation of electron density varies as the square root of $\cos \theta$ [i.e., $(q_m)^{1/2}$]. As shown in Figure 7.3, a reasonably good approximation of a real electron density profile can be obtained by a judicious mix of different Chapman layers.

7.3 Ionospheric Chemistry and the Chapman Model (E-Region)

In order to compare our simple Chapman theory with observations, we need to develop some additional ideas about atmospheric equilibrium photochemistry.

Photochemistry is concerned with chemical reactions that are initiated by the absorption of radiation. Thus schematically we write

$$A + h\nu \rightarrow A^* \qquad (7.19)$$

meaning that molecule (or atom) A absorbs a quantum of energy $h\nu$ and causes an energy transition to the excited molecular (or atomic) state denoted by A^*. For molecules, the excited state could be an excited rotational or vibrational

Table 7.1. Ionospheric processes *(after Chamberlain, 1978).*

Region	Nominal height of layer peak (km)	$N_e^{(max)}$ (cm^{-3})	α_{eff} (cm^3/sec)	Ion production	Recombination
D	90 Lower following solar flare	1.5×10^4 (noon): absent at night	3×10^{-8}	Ionization by solar x-rays, or Lyman alpha ionization of NO. Enhanced ionization following solar flares due to x-ray ionization of all species. Electron attachment to O and O_2 forms negative ions; ratio of negative ions to electrons increases with depth and at night.	Electrons form negative ions which are destroyed by photodetachment (daytime only), associative detachment ($O + O^- \rightarrow O_2 + e$), and mutual neutralization ($O^- + A^+ \rightarrow O + A$).
E	110	1.5×10^5 (noon): $<1 \times 10^4$ (night)	10^{-8}	Ionization of O_2 may occur directly by absorption in the first ionization continuum ($h\nu > 12.0$ eV). Coronal x-rays also contribute, ionizing O, O_2, and N_2. Nighttime E and sporadic E (thin patches of extra ionization) are due to electron and meteor bombardment. Some sporadic E radio reflections may be due to turbulence in normal E layer.	Dissociative recombination $O_2^+ + e \rightarrow O + O$ and $NO^+ + e \rightarrow N + O$.
F1	200	2.5×10^5 (noon): absent at night	7×10^{-9}	Ionization of O by Lyman "continuum" or by emission lines of He. This ionization probably accompanied by N_2 ionization, which disappears rapidly after sunset.	O^+ ions readily transfer charge to N_2 and to O_2. Most of the ionization is thus in molecular form and disappears by dissociative recombination.
F2	300 Height and electron density highly variable. Large daily, seasonal, and sunspot-cycle variations are combined with general erratic behavior.	10^6 (noon); 10^5 (midnight)	$10^{-10} - 10^{-9}$ Variable; probably decreases with increasing height	Ionization of O by same process producing F1;F2 formed because α_{eff} decreases with increasing height; F2 region produces little attenuation of radiation. Additional ionization processes may contribute in F2 that are attenuated in F1.	Recombination of molecular ions as in F1; but limiting process here is charge transfer, giving an attachment-like recombination law.

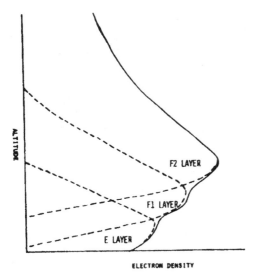

Figure 7.3 Electron density profile as the sum of the E, F1, and F2 Chapman layer contributions.

level, an excited electronic state, or it could be a dissociated or an ionized molecule. For atoms, the excited state could be an excited electronic state or an ionized atom.

Excited molecules (or atoms) can then enter into secondary chemical reactions forming species that might not exist in a state of thermodynamic equilibrium at the local kinetic energy (temperatures) of the reacting molecules. Thermal equilibrium implies that the temperature is the same in all portions of the equilibrium system. Since photochemical reactions are not thermal equilibrium processes, the powerful mathematical methods of thermodynamics are of limited use and, therefore, photochemical processes have to be handled quantitatively using the rate at which each individual reaction occurs.

Within the E-region (90–140 km) ionization is produced mainly from O_2 molecules by photoionization

$$O_2 + h\nu \rightarrow O_2^+ + e, \quad \text{at } \lambda < 1027 \text{ Å} \quad (7.20)$$

which means that a quantum of energy at wavelengths less than 1027 angstroms ($1\text{Å} = 10^{-10}$m) ionizes neutral diatomic oxygen (releasing an electron, e). Remember, in terms of wavelength the photon energy $h\nu$ can be written as hc/λ where c is the speed of light.

Additional ionization of nitrogen molecules N_2 by coronal x-rays produces O_2^+ and NO^+ ions by rapid charge exchange, such as

$$N_2 + h\nu \rightarrow N_2^+ + e, \quad \text{at } \lambda < 796 \text{ Å} \quad (7.21)$$

$$N_2^+ + O \rightarrow N + NO^+, \quad \text{at the rate } k = 5 \times 10^{-10} \text{ cm}^3/\text{sec}$$

$$N_2^+ + O_2 \rightarrow N_2 + O_2^+, \quad \text{at the rate } k = 1 \times 10^{-10} \text{ cm}^3/\text{sec}$$

That is, ion-neutral (or ion-molecule) reactions can lead to the exchange of charge between these two species during a

collision. Notice also that the reaction rates k of equation 7.21 depend on the number density (cm^3) of the neutrals. The production of NO^+ is given by

$$\frac{d}{dt}[NO^+] = [N_2^+][O]k \quad (7.22)$$

where the square bracket is used to denote number density in molecules/cm^3 and k is the reaction rate.

At E-region altitudes, even though CO_2 has reasonably high density, it isn't included with equation 7.21 because of its short chemical lifetime. The charge exchange equations for $N_2^+ + CO_2$ is

$$N_2^+ + CO_2 \rightarrow N_2 + CO_2^+, \\ \text{at a rate } k = 9 \times 10^{-10} \text{ cm}^3/\text{sec} \quad (7.23)$$

but then CO_2^+ interacts quickly with O_2

$$CO_2^+ + O_2 \rightarrow CO_2 + O_2^+, \\ \text{at rate } k = 1 \times 10^{-10} \text{ cm}^3/\text{sec}. \quad (7.24)$$

This brings out the important point that atmospheric chemistry often consists of a series of interlacing reactions which must be computed in a self-consistent manner in order to arrive at physically meaningful results. Table 7.2 lists some of the important charge transfer and ion-atom interchange reactions which are important for planetary atmospheres.

E-region ion production is balanced by the loss of ionization principally by dissociative recombinations such as

$$\left.\begin{array}{l} O_2^+ + e \rightarrow O + O \\ NO^+ + e \rightarrow N + O \end{array}\right\} \begin{array}{l} \text{rate} \\ k = 3 \times 10^{-7} \text{ cm}^3/\text{sec}. \end{array} \quad (7.25)$$

Dissociative recombination occurs as a result of a radiationless transition which forms an unstable intermediate molecule that separates into the individual neutral atomic species. Another possible loss process is by radiative recombination which is schematically written as

$$A^+ + e \rightarrow A^* + h\nu. \quad (7.26)$$

Typically, the reaction rate for equation 7.26 is a factor of 100,000 smaller than the dissociate recombination process given in equation 7.25. Therefore, whenever atmospheric molecules are present, dissociative recombination will be, by far, the dominant loss process. Radiative recombination will become an important loss process only at high altitudes where atmospheric molecules are nearly nonexistent. In any real ionosphere, it is the balance of the ion production rate versus the loss rate which ultimately determines the observed ionospheric density. The loss of ionization in the E-region is given by equation 7.25.

Schematically, the loss process is given by

$$\text{Loss} = k[A^+]N_e \quad (7.27)$$

where k is the reaction rate (cm^3/sec), $[A^+]$ is the ion concentration (O_2^+ or NO^+), and N_e is the electron density (concentration). Within any stable ionosphere, the electron and ion densities are approximately equal (on a scale larger

Table 7.2 Ionospheric reaction rates *(after Bauer, 1973).*

Reaction					k (cm³ sec⁻¹)
H^+ + H	→ H	+ H^+			3 (−9)*
H^+ + O	→ H	+ O^+			4 (−10)
H^+ + CO_2	→ H	+ CO_2^+			1 (−10)
He^+ + O_2	→ He	+ O	+ O^+		2 (−9)
He^+ + CO_2	→ O^+	+ CO	+ He		1 (−9)
He^+ + CO	→ C^+	+ O	+ He		2 (−9)
He^+ + N_2	→ He	+ N_2^+			2 (−9)
He^+ + N_2	→ He	+ N	+ N^+		8 (−10)
N^+ + O_2	→ N	+ O_2^+			5 (−10)
O^+ + O_2	→ O	+ O_2^+			2 (−11)
N_2^+ + O_2	→ N_2	+ O_2^+			1 (−10)
N_2^+ + CO	→ N_2	+ CO^+			7 (−11)
N_2^+ + NO	→ N_2	+ NO^+			5 (−10)
N_2^+ + CO_2	→ N_2	+ CO_2^+			9 (−10)
CO^+ + O_2	→ CO	+ O_2^+			2 (−10)
CO^+ + CO_2	→ CO	+ CO_2^+			1 (−9)
CO_2^+ + O_2	→ CO_2	+ O_2^+			1 (−10)
CO_2^+ + H	→ H^+	+ CO_2			1 (−10)
H^+ + CO_2	→ COH^+	+ O			6 (−10)
H_2^+ + H_2	→ H_3^+	+ H			2 (−9)
He^+ + H_2	→ HHe^+	+ H			< (−13)
C^+ + CO_2	→ CO^+	+ CO			2 (−9)
C^+ + O_2	→ CO^+	+ O			1 (−9)
N^+ + O_2	→ NO^+	+ O			5 (−10)
N^+ + CO_2	→ NO_2^+	+ C			2 (−11)
O^+ + N_2	→ NO^+	+ N			1 (−12)
O^+ + CO_2	→ O_2^+	+ CO			1 (−9)
O_2^+ + N	→ NO^+	+ O			2 (−10)
CO_2^+ + H	→ COH^+	+ O			6 (−10)
CO_2^+ + O	→ O_2^+	+ CO			2 (−10)

*(−b) ≡ 10⁻ᵇ i.e. $*(-b) \equiv 10^{-b}$

than 1 cm ~ the Debye length). Otherwise, large electric fields would form which, in turn, would try to restructure the ionosphere into a configuration which eliminates these non-equilibrium electric fields. Therefore, charge neutrality implies

$$[A^+] \approx N_e \qquad (7.28)$$

and equation 7.27 becomes

$$Loss = k N_e^2. \qquad (7.29)$$

The rate of change of the E-region electron density is simply the difference between the ion production and ion loss processes, or in terms of a quantitative expression

$$\frac{dN_e}{dt} = q_\nu - k N_e^2 \qquad (7.30)$$

where q_ν is symbolic of the production processes given by equations 7.20 and 7.21. Notice that equation 7.30 is in the form of the Chapman model electron density equation 7.17. From equation 7.25, k is 3×10^{-7} cm³/sec which essentially accounts for $\alpha_{eff} \sim 10^{-8}$ cm³/sec previously used for the E-region (Table 7.1). In practice, we find that any atmospheric layer that has a loss process proportional to N_e^2 is explained reasonably well by the Chapman model (i.e., the

E-region is basically a Chapman layer). During the day, the lifetime for molecular ions in the E-region is about 10 seconds.

7.4 F1-Layer and the Chapman Model

In the F1-region (140–200 km) the principal ion formed is O^+ by

$$O + h\nu \to O^+ + e, \qquad \text{at } \lambda < 911 \text{ Å} \qquad (7.31)$$

with some contribution from

$$N_2 + h\nu \to N_2^+ + e, \qquad \text{at } \lambda < 976 \text{ Å}. \qquad (7.32)$$

The distinguishing feature between E- and F1-layers is that the atomic ions making up the F1-region must transfer their charge to molecules prior to recombination because radiative recombination

$$O^+ + e \to O + h\nu, \\ \text{at rate } k = 3 \times 10^{-12} \text{ cm}^3/\text{sec} \qquad (7.33)$$

is extremely slow. Whereas, molecular dissociative recombination will occur relatively fast at these altitudes assuming, of course, that molecular ions are produced. Thus, we believe that the F1-region recombinations are a two-step mechanism: atom-ion interchange,

$$\left. \begin{array}{l} O^+ + O_2 \to O_2^+ + O \\ O^+ + N_2 \to NO^+ + N \end{array} \right\} \text{ at rate } k = \begin{cases} 2 \times 10^{-11} \text{ cm}^3/\text{sec} \\ 1 \times 10^{-12} \text{ cm}^3/\text{sec} \end{cases} \qquad (7.34)$$

followed by the dissociative recombination process given by equation 7.25. Therefore, the loss process for the F1-region is again of the form

$$Loss = k [A^+] N_e \qquad (7.35)$$

or following section 7.3

$$Loss \approx k N_e^2. \qquad (7.36)$$

Thus, the F1-region is also explained reasonably well by the Chapman model. The lifetime for O^+ ions at 180 km is about 60 seconds.

7.5 F2-Layer—Non-Chapman Type Layer

Although the title of this section reveals the answer, let's for the sake of argument assume that the F2-layer is a Chapman layer. Therefore, the electron density profile should be given by equation 7.18. However, at the F2-region altitudes (> 200 km), the atmosphere becomes optically thin to ionizing radiation and the second term in equation 7.12 becomes vanishingly small. Furthermore, atom-ion interchange (equation 7.34) becomes totally dominant at high altitudes and the loss process becomes

$$Loss = k [O^+][A] \qquad (7.37)$$

where A represents N_2 and O_2. Assuming charge neutrality ($[O^+] \approx N_e$), the loss process becomes

$$Loss \approx k\,[A]\,N_e \qquad (7.38)$$

where now the loss process depends only as the first power of N_e. Thus, for the F2-layer, production minus the loss becomes

$$\frac{dN_e}{dt} =$$

$$K_\nu [O]_o\,F_\infty \exp\left(\frac{-(z-z_o)}{H(O)}\right) - k\,N_e\,[A]_o \exp\left(\frac{-(z-z_o)}{H(A)}\right)$$

$$(7.39)$$

where $[O]_o$ and $[A]_o$ are the densities at z_o, and $H(O)$ is the atomic oxygen scale height and $H(A)$ is the scale height for species A. In equation 7.39 we've assumed that species A varies exponentially with height (i.e., diffusive equilibrium). Solving equation 7.39 gives

$$N_e(z) = const \cdot \exp\left\{(z-z_o)\left[\frac{1}{H(A)} - \frac{1}{H(O)}\right]\right\}. \qquad (7.40)$$

Remember, the scale height is simply

$$H = \frac{kT}{mg} \qquad (7.41)$$

and thus equation 7.40 becomes

$$N_e(z) = const \cdot \exp\left\{\frac{(z-z_o)g}{kT}[M(A) - M(O)]\right\} \qquad (7.42)$$

where $M(A)$ is the mass of N_2 and O_2 and $M(O)$ is the mass of atomic oxygen. Examining equation 7.42 we find that the F2-layer electron density (in chemical equilibrium) increases indefinitely with height—obviously an unrealistic model! Thus far, we have neglected atmospheric dynamics which is competing with ionospheric equilibrium chemistry. In fact, within the F-region the motion of atmospheric gases may significantly influence or even dominate ionospheric chemistry.

We will start by considering what happens to the electron concentration when we allow for the vertical movement of electrons. If we assume that the electrons are drifting upward at a velocity v (i.e., moving up into a region of lower pressure), then the change in the number of electrons within the idealized box shown in Figure 7.4 is simply the difference of the number entering the box minus the number leaving the box. That is

$$\frac{\partial N_e}{\partial t} = \frac{(N_{e_1}v_1 - N_{e_2}v_2)}{z_1 - z_2} = -\frac{\partial}{\partial z}(N_e v) \qquad (7.43)$$

where N_{e_1} and N_{e_2} are the electron number densities at altitudes z_1 and z_2 respectively. Similarly v_1 and v_2 are the drift velocities at the respective altitudes. Equation 7.43 is just a one-dimensional continuity (conservation of particles) equation. Since the electrons are moving upward into a

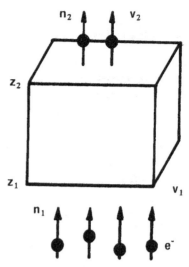

Figure 7.4 Idealized diffusion of electrons through parcel of atmosphere of height z_2–z_1. Electrons enter "box" with velocity v_1 and density N_{e_1} and leave with v_2 and N_{e_2}.

region of lower pressure, their net upward velocity is proportional to the number density gradient. It is customary to write this diffusion drift speed in terms of a diffusion coefficient (D) or

$$v = -\frac{D}{N_e}\frac{\partial}{\partial z}N_e \qquad (7.44)$$

where N_e is again the electron number density. An expression for the diffusion coefficient can be obtained by equating the pressure gradient (assume ideal gas) to the drag force due to collisions. That is,

$$kT\frac{\partial}{\partial z}N_e = -N_e v m\nu \qquad (7.45)$$

where the term on the left is the pressure gradient assuming that the temperature varies only slowly (negligible) with height—an often poor assumption. The right-hand term represents the drag force where N_e is the number density, v is the drift velocity, m is the mass, and ν is the collision frequency (number of collisions per second). Comparing equations 7.45 and 7.44 we see that the diffusion coefficient for this ideal system is

$$D = \frac{kT}{m\nu} \qquad (7.46)$$

and the time constant for diffusion is

$$\tau \sim \frac{D}{v^2} \sim \frac{H^2}{D} \sim \frac{H}{v} \qquad (7.47)$$

where H is the scale height for the electron gas. Remember that up to this point we have assumed an isothermal ionosphere (see equation 7.45). If we wish to make the model more realistic and include temperature, as well as density, gradients then the scale height of the system becomes

$$H^{-1} = \frac{mg}{kT} + \frac{1}{T}\frac{dT}{dz}. \qquad (7.48)$$

However electrons don't act independently of the ions; an electric field exists between them which limits the upward mobility of the electrons and pulls the ions upward. The result of this electrostatic interaction is to produce the *same* scale height for *both* the electrons and the ions and the new scale height is twice the usual (equation 7.41) ion scale height

$$H^*_{electron} = \frac{2kT}{M_i g} \qquad (7.49)$$

where M_i is the ion mass. Therefore, for an ion-electron gas the diffusive time constant is

$$\tau \sim \frac{D}{v^2} \sim \frac{H^{*2}}{D} \sim \frac{H^*}{v} \qquad (7.50)$$

Above the F2-layer peak the atmospheric density is so low that chemical recombinations become less important, but at some lower altitudes chemical recombination and transport are equally important resulting in the F2-layer maximum. Calculation of this peak amounts to a comparison of time constants (or characteristic times) on the principle that of two competing processes the one that occurs fastest will dominate the loss process. The diffusive time constant is given in equation 7.50. The recombination time constant is relatively straightforward. If the source of ionization is removed, the electron density will decay at a rate (using equation 7.38)

$$\frac{dN_e}{dt} = -k[A]N_e. \qquad (7.51)$$

It is customary to rewrite 7.51 as

$$\frac{dN_e}{dt} = -\beta N_e \qquad (7.52)$$

where $\beta = k[A]$ and the process is called Bradbury or β recombination, to distinguish it from the conventional α recombination of equation 7.17. Therefore the time constant for chemical recombination is

$$\tau_c \sim \frac{1}{\beta} \qquad (7.53)$$

where β is a decreasing function of height because the neutral number density, [A], decreases with height (i.e., τ_c increases with height). Equating equations 7.50 and 7.53 places the F2 maximum approximately where

$$\beta \sim \frac{D}{H^{*2}}. \qquad (7.54)$$

Above the F2-peak, the diffusive time constant becomes smaller than the chemical time constant and therefore, the topside F2-layer approaches diffusive equilibrium with a scale height appropriate to the electron-ion gas.

At even higher altitudes O^+ becomes less important and H^+ begins to dominate. It is customary to call the region

dominated by O^+ the topside ionosphere, and the region dominated by H^+ the plasmasphere.

Before we end this section, we need to clear up a point about Figure 7.3. Earlier we said that a reasonable approximation for an electron density profile results from a mix of Chapman layers, even for the F2-layer. This contradicts what we've said above. The only way a Chapman layer works for the F2-layer is to modify equation 7.18 by replacing the $\frac{1}{2}$ in the exponent by 1. The resulting modified Chapman layer works reasonably well near the peak but often fails miserably above the peak. To correct this deficiency, an exponential topside is used to replace the upper portion of the modified Chapman layer. We go through this process because it is simpler computationally (especially for operational applications) to use the modified Chapman model with all its limitations rather than solve the differential equations of this section numerically.

7.6 D-Layer Ion Chemistry

The ionospheric D-region (50–95 km) is characterized by small ionization densities, and large collision frequencies of electrons and ions with neutral molecules. During normal daytime conditions, free electron densities increase nearly exponentially from about 1 cm^{-3} at 50 km to 10^4 cm^{-3} at 90 km. The distinguishing feature of the D-region is the predominance of negative ions. For example, the N_2 molecule does not form a stable negative ion, but O_2 does.

Negative ions are formed by three-body attachment

$$O_2 + e + O_2 \rightarrow O_2^- + O_2$$
$$\text{at rate } k = 5 \times 10^{-31} \text{ cm}^6/\text{sec.} \qquad (7.55)$$

The third body may also be N_2, but in this case

$$O_2 + e + N_2 \rightarrow O_2^- + N_2 \qquad (7.56)$$

and the rate coefficient is much lower. The three-body process is more important at the high D-region densities ($[O_2] \sim 10^{13}$ to 10^{14} cm^{-3}) than radiative attachments

$$O_2 + e \rightarrow O_2^- + h\nu \quad \text{at rate } k \sim 10^{-19} \text{ cm}^3/\text{sec.} \quad (7.57)$$

The loss process consists of removing electrons from O_2^- in the daytime by photodetachment

$$O_2^- + h\nu \rightarrow O_2 + e \qquad \text{at visible wavelengths} \quad (7.58)$$

and by collisional detachment (both day and night)

$$O_2^- + O_2 \rightarrow O_2 + e + O_2$$
$$\text{at rate } k \sim 10^{-22} \text{ cm}^3/\text{sec.} \qquad (7.59)$$

The chemical rate of equation 7.59 is extremely sensitive to temperature and the value used in equation 7.59 is an estimate for 200 K.

Positive ions and free electrons are created by two processes: (1) photoionization of N_2 and O_2 by solar x-rays; and

(2) solar Lyman alpha ionization of the minor atmospheric constituent NO. Symbolically,

$$A + h\nu \rightarrow A^+ + e \qquad (7.60)$$

where A is used to represent N_2, O_2, or NO. The D-region chemistry stops when the positive ions are neutralized, not simply when the free electrons disappear by attachment. The principal recombination process is mutual neutralization

$$A^+ + B^- \rightarrow A + B$$
$$\text{at rate } k \sim 2 \times 10^{-7} \text{ cm}^3/\text{sec}. \qquad (7.61)$$

Electrons can also disappear through dissociative recombination (equation 7.25), a process which must be considered for any daytime equilibrium calculations.

For many years it seemed that the O_2^- chemistry summarized above accounted for the D-region ionization in a satisfactory way. However, equation 7.58 detaches electrons with visual light, a prediction not supported by observations. Experiments indicate that electrons are bound to molecules which require ultraviolet rather than visual light to photodetach the electrons. Laboratory work suggests that NO_2^- and NO_3^- ions may be important to the D-region chemistry.

The ion chemistry is further complicated by hydration; the large dipole moment of water molecules allows it to become attached to both positive and negative ions. These loose associations can affect both the photodetachment frequency and reaction rates of the molecular ion as compared to a nonhydrated ion. Furthermore, due to the small degree of D-region ionization and relatively large neutral air densities, transport of minor constituents into or from the D-region may significantly influence the ionization balance.

As mentioned above, positive ions are produced by Lyman alpha radiation in the upper portions of the D-region by ionization of NO (nitric oxide). However, cosmic rays and solar x-rays (especially by flares) penetrate deeper into the D-region and presumably ionize O_2 and N_2 which will then readily transfer positive charge to NO during collisions. Recently, mass spectrometers flown on rockets have shown water-derived ions such as H_3O^+, $H_5O_2^+$, and $H_7O_3^+$ may also be important.

7.7 Summary

Although ionization occurs throughout the atmosphere, it is most effective above about 75 km where a significant concentration of free electrons (and ions) persists for extended periods of time. The electron density at any altitude is determined by the relative ratio of recombination and ionization at that altitude. The ionosphere has a number of distinct peaks because the atmospheric composition and density change with altitude. These layers, in order of increasing peak density and altitude, are called the D, E, F1 and F2 regions. The D- and F1-layers disappear completely at night,

and the nighttime E becomes so weak that, normally, its presence is of no consequence. However, the F2-layer does persist throughout the night, although its intensity continuously diminishes, reaching minimum just before sunrise. The ions and electrons making up the ionosphere are only a minor atmospheric constituent making up only about 0.4% of the total atmospheric material at a given altitude.

7.8 References

Appleton, E.V., and Barnett, M.A.F. 1925. On Some Direct Evidence for Downward Atmospheric Reflection of Electric Waves, in *Proceedings of the Royal Society*, London, A109, p. 621.

Bauer, S. J. 1973. *Physics of Planetary Ionospheres*. Springer-Verlag, New York.

Bauer, P. 1981. Thermospheric Neutral Composition Changes and Their Causes, in *AGARD Conference Proceedings*, No. 295. *The Physical Basis of the Ionosphere in the Solar-Terrestrial System*. Harford House, London.

Bostrom, R. 1973. Electrodynamics of the Ionosphere, in *Cosmical Geophysics*, Egeland, Holter, and Omholt, eds. Universitetsforlaget, Oslo, Sweden.

Chamberlain, J. W. 1978. *Theory of Planetary Atmospheres; An Introduction to Their Physics and Chemistry*. Academic Press, New York.

Hanson, W. B. and Carlson, H. C. 1977. The Ionosphere, in *The Upper Atmosphere and Magnetosphere*. National Academy of Sciences, Washington, D.C.

Hargreaves, J. K. 1979. *The Upper Atmosphere and Solar-Terrestrial Relations*. Van Nostrand Reinhold Co., New York.

Moyer, V. 1976. *Physical Meteorology*. Texas A & M University, College Station, Tex.

Nisbet, J. S. 1975. Models of the Ionosphere, in *Atmospheres of Earth and the Planets*, B. M. McCormac, ed. D. Reidel Pub. Co., Boston.

Rawer, K. 1952. Calculations of Sky-wave Field Strength. *Wireless Engineer*, vol. 29, p. 287.

Rush, C. 1979. Report of the Mid- and Low-Latitude E and F Region Working Group, in *Solar Terrestrial Predictions Proceedings*. U.S. Government Printing Office, Washington, D.C.

Smith-Rose, R.L., and Barfield, R.H. 1927. Further Measurements on Wireless Waves Received from the Upper Atmosphere, in *Proceedings of the Royal Society*, London, A116, p. 682.

Tascione, T. F., Flattery, T. W., Patterson, V. G., Secan, J. A., and Taylor, J. W. 1979. Ionospheric Modeling at Air Force Global Weather Central, in *Solar Terrestrial Predictions Proceedings*. U.S. Department of Commerce, U.S. Government Printing Office, Washington, D.C.

7.9 Problems

7-1. What is the scale height for a plasma in a gravitational field when there are two ions with density, mass, and temperature of N_1, M_1, T_1, and N_2, M_2, T_2? Assume that $N_e = N_i$, $T_1 = T_2 = T_i$, and that $T_e \neq T_i$.

7-2. Using the results from problem 7-1, show in the special case where $N_1 \gg N_2$ and $T_e = T_i$, that the plasma has twice the scale height of the parent neutral of mass M_1.

7-3. Starting with the definitions given in section 7.2, derive the Chapman layer equation 7.18. Some ionospheric models try to use a modified Chapman layer approximation for the F2-region by changing the 1/2 in the exponential to a 1. Physically, what does this change correspond to? Is it a reasonable approximation?

7-4. What is the difference between β and α recombination chemistry? Where does each type of recombination chemistry dominate and why?

7-5. Why does D-region chemistry differ from the chemistry of the other ionospheric layers? Would the D-region have any significant impact in radio wave propagation? Why?

7-6. Distinguish between dissociative recombination and radiative recombination. Where is each type of recombination important and why?

Chapter 8

Ionospheric Variability

8.0 Introduction

In Chapter 7 we described what might be called the "ideal" ionosphere. As you can imagine, the ionosphere rarely matches this ideal model and the variation from ideal can range from minor to extreme. In fact, some fluctuations of the ionospheric electron density are regarded as "normal." For example, in Chapter 7 we noted that at night the D-region and the distinction between the two daytime F-regions (F1 and F2) disappear. Also at night, there is a marked decrease in the maximum electron densities in the E- and F2-regions by one or two orders of magnitude. The daytime ionizations in the E- and F1-regions are larger in summer than in winter, but the reverse is often true for the F2-region. This is called the F2 seasonal anomaly; it dates back to the time when ionospheric physicists held the Chapman model to be valid for the entire ionosphere and departures from expected Chapman behavior were termed anomalies. Thus the names diurnal anomaly (F2-region maximum usually is not at noon but rather between 1300 to 1500 hours local time) and equatorial anomaly (the F2-region peak departs from the expected solar zenith angle dependence within $\pm 20°$–$30°$ of the geomagnetic equator) have become standard ionospheric terminology. Also in Chapter 7, we showed that there is a definite relationship between solar activity and electron density. At sunspot minimum, the electron densities are lower by a factor of two to four than at sunspot maximum especially in the F-region.

Some other regular variations in electron density in the ionosphere result from atmospheric tides produced by the Sun and Moon. These tides, whether due to purely gravitational effects or to daytime heating, produce vertical swelling of the entire atmosphere, including the ionosphere. The movement of the conducting ionosphere, in the presence of a magnetic field, causes electric currents to flow as a result of the familiar dynamo action. Thus, tides produce a regular, although minor, change in the structure of the ionosphere.

In the following sections, we will discuss "non-normal" variations to the ionosphere. In particular we will discuss the auroral storm, sudden ionospheric disturbances, and traveling ionospheric disturbances. We will also discuss phenomena unique to particular latitude belts of the ionosphere.

8.1 Ionospheric Disturbances

A sudden ionospheric disturbance (SID) is much like the sudden commencement storm in the magnetosphere. The SID manifests itself within a few minutes of the appearance of some strong solar flares as a sharp fadeout of long-distance, radio communication on the sunlit side of the Earth. The short-wave fadeout (SWF) is caused by strong enhancement of the electron density of the D- and lower E-regions, as a result of the penetration of solar x-rays to these low levels. High-frequency radio waves, that normally would pass through the D-region and be reflected at higher levels, are absorbed instead (see Chapter 9). The disruption of communications may last an hour or so.

The period of fadeout corresponds roughly to the duration of the more active phase of the solar flare plus the time required for the additional atmospheric ionization to recombine. The absence of SID effects transported to the dark side of the Earth is due largely to the rapid recombination of electrons and ions beyond the terminator.

In general, *ionospheric storms* are characterized by a more gradual onset and longer duration than in the case of SID. Some persist for several days and cover large portions of the globe. Two storm types of particular interest are PCAs and geomagnetically induced storms.

The polar-cap absorption (PCA), as the name implies, is observed only at high latitudes, where it is accompanied by a communications blackout resulting from an increase in electron density at altitudes between 55 and 90 km. The effect is associated with arrival of very energetic solar flare protons (<10 MeV). These particles are guided by the Earth's magnetic field lines (cusp) directly into the polar caps. Here, they penetrate to altitudes as low as 50 km before giving up their energy in ionizing neutral atmospheric constituents. Since flare-related protons are characterized by a considerable spectrum of energies, the most energetic protons arrive first, producing small patches of ionization, and as the bulk of lower energy protons arrive, the PCA gradually expands to fill the polar cap. Absorption will usually be most intense in the sunlit polar cap. In contrast to the SID, the PCA is a long-lived effect with durations ranging from tens of hours to several days.

The second type of ionospheric storm is the *geomagnetically induced* storm. It also is associated with increased electron densities in the lower ionosphere. Storms start about 20 hours after some solar flares, and they may persist for some time beyond the end of the geomagnetic disturbance. Unlike PCA events, they are more intense at night. Although storm related effects are most frequent in the auroral zones centered at ±67° geomagnetic, the absorption of radio waves

is also enhanced at lower latitudes. Within the ionosphere the auroral substorm begins with a sudden brightening of one of the quiet auroral arcs or a sudden formation of an arc, followed by a rapid poleward motion, resulting in an "auroral bulge" around the midnight sector. As the auroral substorm progresses, the bulge expands in all directions; the evening sector is characterized by a "westward traveling surge" and in the morning sector of the bulge, the arcs disintegrate into

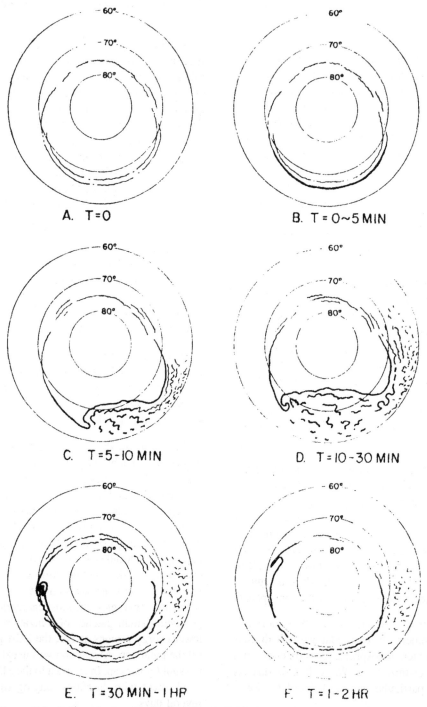

Figure 8.1 Schematic showing details of an individual auroral substorm *(after Akasofu, 1968)*.

patches which drift eastward at speeds about 300 m/s. The expansive phase ends and the recovery phase starts when the expanded bulge begins to contract (see Figure 8.1).

For many years, ground observers have seen ionospheric disturbances travel from high latitudes toward the equator. These traveling ionospheric disturbances (TIDs) appear to be produced by auroral heating events which generate atmospheric gravity waves; as these gravity waves move equatorward they modify the mid-latitude ionosphere (see section 6.6). Near the active auroral regions, large upwellings of heated gas from the lower thermosphere drastically alter both the neutral and ion compositions at higher altitudes. The disturbances propagate as waves in the neutral atmosphere that can be detected through their effects on the ions.

8.2 Ionospheric Storm Morphology

Considerable energy is injected into the auroral oval during a geomagnetic disturbance (see Figure 8.2). Approximately a third of the energy is due to precipitating particles. The remainder is due to joule heating by storm-associated currents. This energy, injected near the turbopause (105 km), influences the chemical composition of the thermosphere. Molecular nitrogen and oxygen bubble up into the F-region and decrease the available atomic oxygen, a serious consequence since atomic oxygen is a primary source of photoelectrons. Thus, N_2 and O_2 (by way of NO^+, O_2^+, and N_2^+) combine with available free electrons and reduce the ambient electron density. Neutral winds then displace these compositional changes to lower latitudes (in the summer hemisphere and at night) or constrain them to the higher latitudes (during winter and in the sunlit hemisphere).

At mid- and high-latitudes, a plot showing the percent change of the ionospheric maximum electron density during an ionospheric storm exhibits the same general pattern as a magnetometer trace during a magnetospheric storm (see Figure 8.3). During the first few hours of an ionospheric storm, the mid- and high-latitude electron density and total electron content (TEC) may increase 10% over normal values. Subsequently, both the maximum density and TEC decrease to some minimum value and then slowly recover to normal values in a few days. Analogous to magnetospheric storms, these three phases are often referred to as the initial, main, and recovery phases. Some authors refer to the enhanced phase as the "positive phase" and the remainder of the storm as the "negative phase". Ionospheric storms vary greatly from one another: the start of a storm may be very sudden or in other cases the start may be very gradual. In addition, ionospheric storms which start in the nighttime hours generally do not show any positive phase.

During the main phase of the storm, the height of the F2 peak appears to rise in ionosonde traces. An ionosonde uses a radio-echo technique to determine the time it takes for a signal of known frequency to be reflected by the ionosphere; the round trip time is used to calculate the height of the reflecting layer. The round trip time depends on the electron density distribution because the radio wave's group velocity slows as the electron density increases (see section 9.2). In fact, careful analysis of ionosonde measurements show that most of the apparent height increase is due to the thickening of the peak region which slows the radio signal more than expected for normal ionospheric conditions.

The effects of an ionospheric storm can be particularly insidious at some locations. Near the auroral oval, a station may see normal conditions during quiet times and again during very disturbed times. Conversely, a small disturbance may result in a severe depletion of the overhead electron density. A slightly stronger disturbance may markedly in-

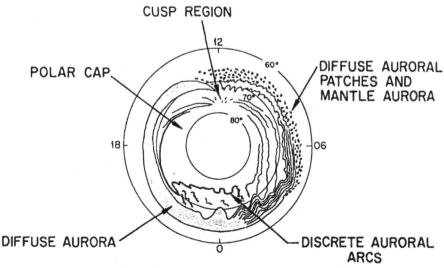

Figure 8.2 An illustration of the principal phenomenological features in the high-latitude auroral zone during geomagnetic substorms (after Akasofu, 1981).

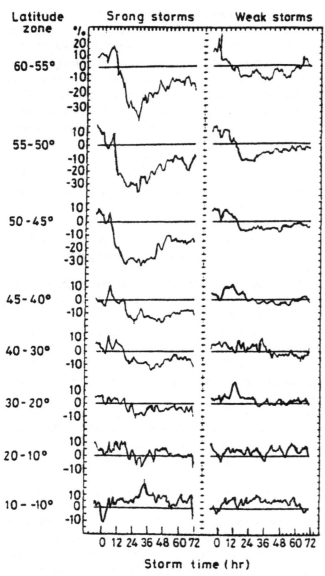

Figure 8.3 Variation of the foF2 electron density as a function of geomagnetic latitude. The percentage change shown on the vertical axis refers to the deviation from quiet day behavior *(after Matsushita, 1959)*.

crease overhead densities. The culprit here is the shifting position of the auroral oval with level of geomagnetic activity. A minor disturbance may move the oval just far enough equatorward to place our imaginary station under the auroral oval, whereas a major storm may move the auroral zone so far equatorward that the polar ionosphere may extend over the station. Since the bottom portion of the polar ionosphere is essentially unaffected by a geomagnetic disturbance, only the normal polar irregularities would differentiate this condition from the quiet ionosphere normally recorded at the site.

Low-latitude sites may also experience surprising results from a geomagnetic disturbance. Weakening of the high-altitude geomagnetic field by the storm ring current slows the vertical transport at the equator (see section 8.3). This

permits the equatorial regions to retain a larger portion of the electron density produced there. This enhancement gradually dissipates as the disturbance subsides, and vertical transport returns to pre-storm levels.

8.3 Low-Latitude Phenomena

At the equator, the magnetic field is horizontal, and this special geometry leads (in the E-region) to an intense current sheet, known as the equatorial electrojet (not to be confused with magnetospheric ring current). This electrojet flows along the magnetic equator at an altitude of about 100 km and is concentrated in a strip only a few degrees wide in latitude. The current flows toward the east by day and the west by night, but the westward currents are nearly undetectable in magnetometer measurements because of the small electron concentrations at night (see section 4.4).

The electric fields associated with ionospheric currents generally drive a plasma convection in the F-region at low magnetic latitudes that is upward and westward in the day-time, and downward and eastward at night. The upward motion in the daytime raises freshly ionized plasma near the equator to great heights, where recombination is slow. Subsequent diffusion, or flow, down the magnetic field lines under action of gravity adds this extra plasma to that produced locally at higher latitudes. The result of this plasma transport, illustrated in Figure 8.4, is that ionization peaks are formed in the subtropics on each side of the magnetic equator (i.e., the "equatorial anomaly" discussed in section 8.0). This phenomenon is often referred to as the fountain effect. The latitude of the peak formations is often not symmetrical about the magnetic equator because the plasma transport along the magnetic field lines can interact with the neutral winds. The neutral winds usually cause plasma to be pushed from the summer to winter hemispheres near midday, so that winter hemisphere peaks are larger. However, since the magnetic equator is tilted with respect to the geographic equator, longitudinal anomalies occur in this seasonal behavior which can affect the magnitude and even the direction of the wind component along the field lines.

The tilt of the geomagnetic field results in anomalous areas in Southeast Asia and over the South Atlantic. As seen in Figure 9.4, the offset of the geomagnetic field from the geographic center results in the radiation belts being closer to the Earth's surface in the South Atlantic region. This means as the trapped particles gyrate about the magnetic field, their gyroradius carries them to lower altitudes (relative to the Earth's surface) where they have a higher probability of colliding with neutrals. Since trapped electrons drift eastward, they interact with neutrals mainly along the western edge of the South Atlantic anomaly. Protons, drifting westward, interact along the eastern anomaly boundary. This interaction tends to accentuate F-region anomalies (and, if energies are sufficient, may increase D-region absorption).

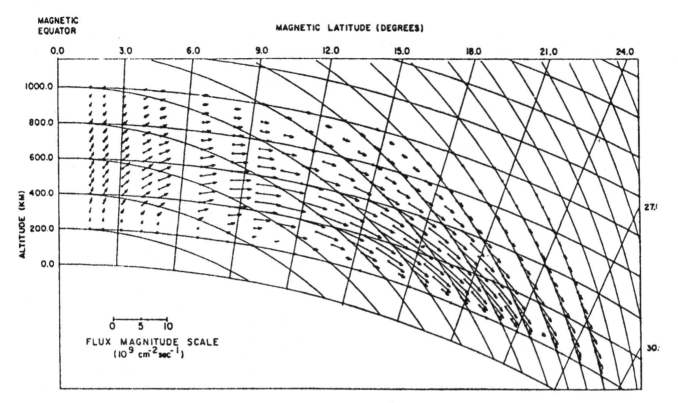

Figure 8.4 Plasma drift along magnetic field lines leading to the fountain effect *(after Townsend, 1982).*

The Southeast Asian anomaly is the reverse of the Atlantic situation, with stronger magnetic field strengths present at the same altitude as in the South Atlantic anomaly. This means that anything (e.g., a geomagnetic disturbance) which disturbs the trapping regions may release higher energy particles into the Asian area. D-region absorption is a likely result of particle dumping in the Southeast Asian area. As with many other portions of the ionosphere, these anomalies are still not well monitored. Additional work in both monitoring and theory is needed.

Turbulence in the ionosphere is accompanied by rapid changes of electron density in both time and space that may last a few minutes to several hours. At F-region altitudes this results in spread F, so called because ionosonde measurements indicate a thicker than normal F-region; therefore, it can be said that the region is "spread" in depth. Equatorial spread F is almost exclusively a nighttime phenomenon that is more prevalent near the equinoxes. Communication satellites have revealed that equatorial spread F can seriously distort radio waves even in the gigahertz frequency range (see section 9.13). The total electron concentration, and hence the index of refraction, is very irregular with changes in concentration as large as a factor of 10^2 or 10^3 in only a few kilometers. Scale sizes as small as 60 m have been observed from Atmosphere Explorer satellites. The ionospheric research radar at Jicamarca, Peru, extends this limit down to 3 m. Not only are the radar echoes enhanced by up to 60 or 70 dB during spread F, but these abnormal signals can

appear in a time of less than 8 msec over regions tens of kilometers across. Sometimes the abnormal echoes persist for hours, and during this time the altitude of a particular irregularity patch will often increase by hundreds of kilometers. Detailed examination of the echoes also reveals a frequency spreading that corresponds to a turbulent velocity on the order of several hundred meters per second.

A similar behavior is shown in the ion concentration and mean ion velocity data recorded by the Atmosphere Explorer satellite near the equator at night. Measurements show low ion density regions tend to move upward, as bubbles rise in a denser medium, and indeed it has been suggested that spread F structure is caused by such buoyancy forces.

This buoyancy mechanism can be explained using Figure 8.5. Under certain conditions convecting ionospheric plasma patches are susceptible to the $\vec{E} \times \vec{B}$ gradient drift instability. In many ways, the $\vec{E} \times \vec{B}$ drift instability resembles the Rayleigh-Taylor instability in which a heavy fluid is supported by a lighter fluid. In Figure 8.5 (a), the coordinates are aligned with x and y as shown and the z-axis pointing out of the page. The ambient geomagnetic field ($\vec{B}_o$) is aligned along z, the convection velocity is in the −y direction, the gravitational force is along x, and the electron density gradient increases in the −x direction. Physically, this picture represents the conditions after sunset in the equatorial ionosphere just below the F region. At this time, the D, E, and F1 regions have disappeared, and for all practical purposes we

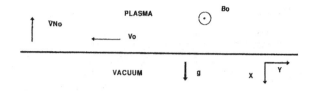

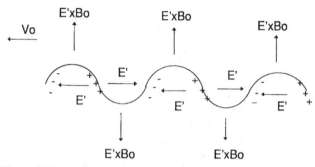

Figure 8.5 (a) Sketch showing the idealized conditions in the low-latitude F-region after sunset (no E- or F1-regions); (b) the physical mechanism of the $\vec{E}' \times \vec{B}$ gradient drift instability (*after Keskinen, 1984*).

can approximate conditions below the F-region as a vacuum. Under steady state conditions, one can show (see problem 8-11) that the drift velocity becomes

$$\vec{V}_o = \frac{M_i}{q} \frac{\vec{g} \times \vec{B}_o}{B_o^2} \qquad (8.1)$$

where M_i is the mass of the ions, g is the gravitational acceleration and q is the charge on the ion. The electrons drift in the opposite direction with a velocity in the form of equation 8.1 except the velocity is proportional to the mass of the electron. Therefore, the electron drift velocity is negligible compared to the ion drift velocity.

If a ripple develops along the interface as the result of random thermal motions, the drift velocity will cause the ripple to grow (see Figure 8.5 (b)). As the ions begin to drift, they cause a charge build up on the sides of the ripple. In addition, an electric field ($\vec{E}'$) develops which changes sign from crest to trough in the perturbation wave. The $\vec{E}' \times \vec{B}_o$ drift is always upward in those regions where the surface has moved upward, and is directed downward wherever the wave has moved downward. As a result, the ripple grows (i.e., goes unstable) because of these phased $\vec{E}' \times \vec{B}_o$ drifts. When the ripples move upward, steep density gradients along the edges can become the source of smaller scale irregularities which produce intense distortion of gigahertz frequency radio waves.

As they rise, these "underdense plumes" spread north and south along geomagnetic field lines. Underdense plumes seem to form 2 hours or less east of the sunset terminator, possibly due to the abrupt reversal of the equatorial electrojet (and associated $\vec{E} \times \vec{B}$ forces) near the terminator. Initially,

they drift rapidly (500 m/sec) westward toward the terminator as they rise and increase to maximum size. They then drift more slowly (100–200 m/sec) eastward. A fully developed plume actually approximates a flattened cylinder along a geomagnetic field line. The largest density discontinuities (by comparison to ambient value) occur between 225 and 450 km altitude, but they may extend to an altitude of 1000 km. The fully developed anomaly is 50 to 300 km thick (vertical), and extends 100 to 1,500 km east-west and 2,000 km or more north-south. It may persist for 2 to 3 hours. Similar structures are observed to form after midnight, often in association with increased geomagnetic activity.

8.4 Midlatitude Phenomena

The midlatitude region is free of the direct influence from phenomena associated with the horizontal magnetic field geometry peculiar to the equatorial region. Also, it is generally free of the direct influence of energetic particle precipitation and large electric fields associated with the auroral zone. However, the energy deposited into the neutral atmosphere at high latitudes can profoundly influence the midlatitude ionosphere by changing the atmospheric circulation patterns that determine the neutral gas composition.

The midlatitude F-region is generally thought to be the best understood region of the ionosphere (see Figure 8.6). The maximum electron concentration occurs at the level where downward diffusion and electron loss by recombination are of comparable importance. At that altitude, the electron concentration is nearly in photochemical equilibrium, with a balance between the photoionization of atomic oxygen and electron recombination with molecular ions (see section 7.5). Variations in the ratio $O/(N_2 + O_2)$ thus can lead to important variations in peak density. Vertical ion drift (due to neutral winds or electric fields) can shift the altitude of the F2-peak and also its concentration. In addition, exchange of ionization along the magnetic field lines between the ionosphere and the plasmasphere (the high-altitude region where H^+ ions predominate) is of importance to the maintenance of the nighttime ionosphere. Appropriate combinations of these various factors have been applied to explain a variety of phenomena initially viewed as "anomalous." Yet there are other gross features of the global morphology of the midlatitude ionosphere that still await basic understanding.

Ideally, the F2-peak electron density should monotonically decrease until its sunrise increase. However, there are periods where there is no appreciable nighttime decrease. In some of these cases, an observed equatorward wind, which has an upward component along the magnetic field, raises the height of the F2-peak at sunset to a level where the molecular concentration and resultant recombination rate are an order of magnitude lower than during the daytime. In fact, the general F-region atmospheric wind pattern tends to blow from the hot daytime subsolar to the cold nighttime antisolar

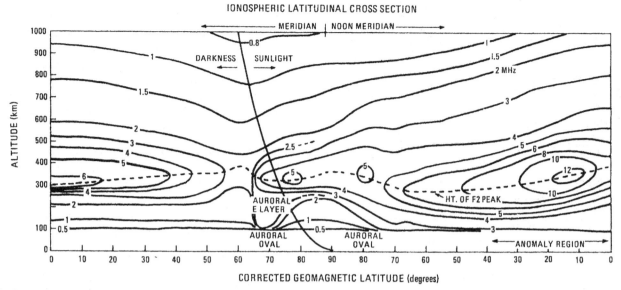

Figure 8.6 Schematic cross section of the ionosphere in the noon-midnight meridian plane (across the magnetic pole). The figure is for minimum solar activity at the equinox (1200 UT). Contour lines of plasma frequency are plotted in units of megahertz (MHz). The conversion to electron density is $N_e = 1.24 \times 10^4 \times (frequency)^2$, where N_e is in units of inverse cubic centimeters and frequency is in MHz.

point, as well as from the summer to the winter hemisphere. This effect, combined with a downward flow of plasma from the plasmaspheric (H^+) reservoir, has been found adequate to maintain the nighttime F-region when tested against the very limited set of measurements.

However, a much more common nighttime evolution includes an abatement or reversal of the equatorward wind in the middle of the night, allowing the plasma to slide down the magnetic field lines to regions of a greater recombination rate. This "midnight collapse" can lower the F2-peak density by an order of magnitude over a time period of approximately an hour and enhance electron concentrations below 200 km by two to three orders of magnitude. This event spans the entire midlatitude sector, although less pronounced and occurring later in time with increasing latitude. This ionospheric effect was discovered over two decades ago by workers in high-frequency communication prediction services. Only now, however, are the measurements becoming available that may identify the driving mechanism for this wind.

There is also a seasonal or winter anomaly in daytime F2-peak density; although the solar radiation is more nearly overhead in local summer, F2-peak electron density achieves significantly greater daytime values in local winter. The winter F-layer anomaly is most apparent in the upper middle (45°–55°) latitudes of the winter hemisphere. Midday electron densities may exceed those in the summer hemisphere by a factor of four. As a consequence, the diurnal variation in the F-layer electron density is a maximum in the winter hemisphere. This effect, a consequence of conjugate (hemisphere) transport, nearly disappears during solar minimum.

Conversely, maximum nighttime F-layer densities are observed in the summer hemisphere because of the longer period of daytime ionization. The effect appears to correlate best with geomagnetic latitude in the northern hemisphere but with geographic latitude in the southern hemisphere.

The F-layer also shows a strong solar cycle variation. Figure 8.7 compares F2 plasma frequency variations to those of the F1- and E-layers for a given upper middle latitude site. Notice that the F2 plasma frequency (and, by inference, electron density) increases steadily up to a sunspot number of about 150 and seems to level slightly for higher levels of solar activity. This probably results from an increase in F-layer collisional recombination due to increased heating which enhances neutral densities at F-layer heights partially offsetting the increased ionization. It is important to remember that the solar sunspot number (SSN) has no direct relationship with F-layer electron density. SSN is taken as a convenient index assumed representative of solar EUV emission. While this is generally true, it does not hold in many specific cases. Even the commonly used F10 index (10.7 cm radioflux emission) is not always representative of ionizing emission levels at a given time. Extreme care must be taken in the interpretation of correlations relating F10 or SSN to ionospheric response (also see section 6.3). Figure 8.8 summarizes the effect of seasonal and solar cycle effects on the ionospheric electron density profile.

8.5 High-Latitude Phenomena

In the mid- and low-latitude E- and F-regions, the primary source of ionization is solar EUV and UV. The day-to-day

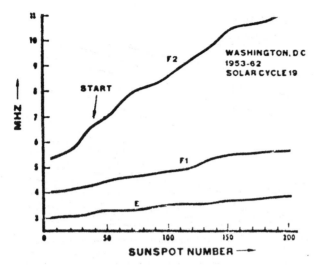

Figure 8.7 Ionospheric response to varying levels of solar activity as measured by sunspot number *(after Townsend, 1982).*

relative contribution of the solar and electron precipitation energy inputs into the high-latitude ionosphere. In fact, the high-latitude ionosphere is not a simple receptor of input energy sources. As discussed in section 5.4, changes in the ionospheric conductivity can provide active current feedback along the plasma sheet magnetic field lines that thread the ionosphere making the high-latitude ionosphere a plasma source for the magnetosphere.

The strong electric fields which arise from the interaction of the interplanetary magnetic field (IMF) and the geomagnetic field drive a two-celled ion convection pattern with anti-sunward ion flow across the pole and a return flow along the auroral oval. At E-region heights, the ion convection is restrained by collision with neutrals.

In Chapter 6, Figure 6.8 shows a schematic of the two cell convection pattern for B_z south conditions. Generally, the dusk convection cell is larger than the dawn cell. In addition, experiments show for B_z south and B_y positive the antisunward flow can be seen above 70° geomagnetic in the dusk sector for large values of K_p; similar flow is not seen in the dawn sector. Conversely, when K_p is large, B_z is south, and B_y is negative, the antisunward flow can be seen in the dawn sector above 70° geomagnetic and not in the dusk sector. The result is a net current flow within the auroral zone called the auroral electrojet (see Figure 4.5). Observations also seem to indicate that under extremely quiet conditions when B_z is

variations in the ionosphere essentially reflect changes in the neutral atmosphere density and winds. Electric fields occasionally become an important modulating force during disturbed periods. At high latitudes, the magnetosphere-ionosphere interaction is the dominant dynamic force. The precipitation of energetic particles is an important source of ionization, particularly in the E-region. Figure 8.9 shows the

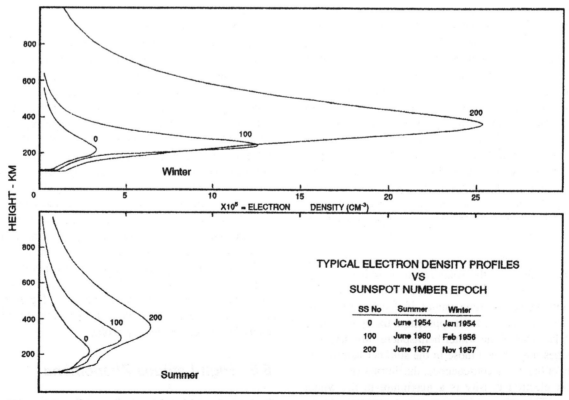

Figure 8.8 Seasonal and solar cycle variation in the noon electron density distribution. The graphs are arranged to show increments of 100 in sunspot number. The electron densities above the peak are extrapolated *(after Wright, 1962).*

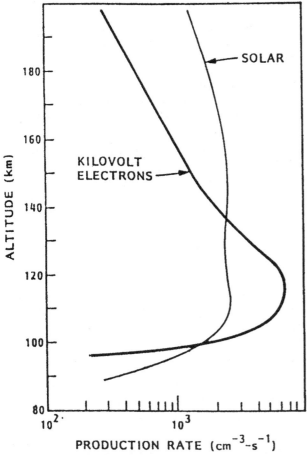

Figure 8.9 Schematic comparing the relative energy input at high-latitudes between precipitating electrons and solar electromagnetic radiation *(after Imhof, 1986)*.

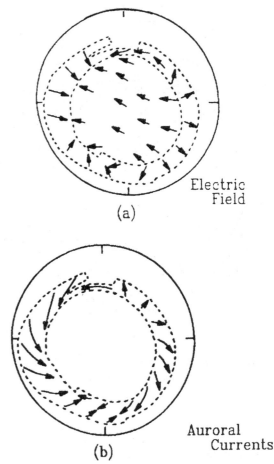

Figure 8.10 (a) Sketch of high-latitude electric field pattern. Dashed lines represent location of auroral oval. (b) Schematic of the resulting horizontal auroral zone current pattern; the polar cap currents are not shown *(after Chiu et al., 1984)*.

north the antisunward flow of plasma across the polar cap may be replaced by two additional cells with flow paths opposite to those in the dawn and dusk sectors.

The general sunward plasma flow in the auroral zone produces an electric field ($\overrightarrow{V} \times \overrightarrow{B}$) directed equatorward in the dawn side and poleward on the dusk side of the oval. This electric field pattern is consistent with the region 1 and 2 Birkeland currents described in Chapter 5. As a reminder, the dawn side region 1 currents flow downward on the polar side of the auroral oval while the region 2 currents flow upward on the equatorward side of the oval. On the dusk side the currents are reversed: upward (region 1) on the poleward side and downward (region 2) on the equatorward side. Panel (a) of Figure 8.10 shows a schematic of the electric field distribution within both the auroral zone (outlined by the dashed lines) and the polar cap. Panel (b) shows the resulting horizontal current pattern in the auroral zone.

Frequent collisions between the convecting ions and the ambient high-latitude neutral atmosphere adds significant (joule) heat energy to the neutrals. In turn, the added heat

energy changes the neutral density and winds which act as a feedback mechanism changing the density and structure of the high-latitude ionosphere. Each of the high-latitude E- and F-region anomalous features is related to the magnetosphere-ionosphere interactions.

High-latitude ionospheric phenomenology is usually divided into the diffuse aurora, discrete auroral arcs, the polar cap, and the polar cusp (see Figure 8.2). The statistical distribution of aurora in geomagnetic latitude and local time coordinates has the shape of an oval belt, located near 75° geomagnetic latitude at noon and approximately 65° geomagnetic latitude near local midnight. Figure 8.11 shows an average oval pattern for $K_p = 2$. Generally, the average oval pattern expands equatorward as K_p increases; during extremely disturbed geomagnetic conditions, the expansion can extend below 60° geomagnetic in the dusk sector. Some observations indicate the oval expansion in the dusk sector is less than 2° for each unit increase in K_p, whereas in the dawn sector the expansion is probably less than 1° for each unit increase in K_p.

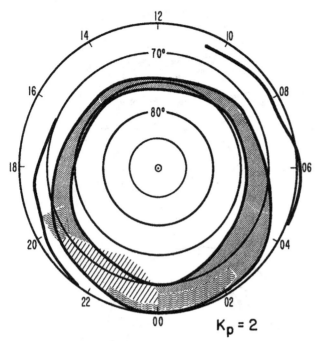

Figure 8.11 Sketch of the average auroral oval shape for $K_p = 2$ conditions; the hatched and wavy lines depict average location of discrete and multiple arc locations *(after Whalen et al., 1985)*.

The most obvious characteristic of the oval is the visible light emissions that are used to demarcate its boundaries. The equatorward edge of the oval marks the boundary between the magnetic field lines that are essentially dipolar, and the field lines that are interacting with the IMF and are being dragged into the magnetotail. Most of the particle precipitation which creates the visible aurora occurs in the nighttime sector. In the day sector, the aurora becomes weaker, with less latitudinal extent, and during very quiet times it is often not seen but never completely disappears.

The aurora seen near the equatorward edge of the oval and most of the daytime aurora is known as the diffuse aurora. It is caused by the precipitation of particles with a relatively broad spectrum of energies (e.g., electrons of 100 eV to 100 keV). Most of the energy results in strong ionization in the E-region, creating a broad shelf in the electron density profile. In addition, the lower energy particles also provide some enhancements to the F-region density. Approximately 5° poleward of the diffuse aurora is the region of discrete auroral arcs produced by monoenergetic field-aligned particle beams. These electron beams are formed by parallel electric fields. As mentioned in Chapter 5, the total potential drop of several kilovolts is probably produced by multiple potential drops along a given set of field lines. Figure 8.2 summarizes ground-based observations which show the relative location of the evening discrete arcs, the morning side multiple arc systems, and patches. Satellite measurements show definite inverted-V structures associated with discrete auroral arcs in the evening sectors. However, the morning-

side aurora do not show such clear signs of inverted-V structures. The evening sector electric field structures support upward field aligned current densities exceeding 10^{-6} A/m².

In Chapter 6, Figure 6.8 shows the location of a satellite orbit relative to an average two cell convection pattern for B_z south. In both the dawn and dusk sectors, the relatively smooth transition from the ion corotation velocities (0.2 km/sec) to significant sunward velocities begins at the equatorward edge of the diffuse aurora. In some cases, the transition may occur equatorward of the optically visible diffuse aurora. The reversal between sunward to antisunward convection takes place over distances of 10 to 200 km and the dusk side reversal is associated with discrete precipitation events. Detailed observations show that a single well-defined reversal often does not exist. Instead, significant changes in electric field direction and magnitude exist at scale sizes down to several hundred meters. The dawnside reversal shows similar characteristics, although electric field structures are somewhat weaker. In addition, the average precipitating particle energy near the dawn reversal is smaller than on the dusk side.

As mentioned above, the auroral oval expands equatorward during geomagnetic disturbances and, with the expansion, the intensity, the number, and strength of discrete arcs also increase. The onset of a substorm is marked by the brightening of the diffuse aurora and the discrete arcs, as well as an equatorward expansion of the diffuse auroral boundary. The westward traveling surge develops from the initial brightening on the northern edge of the oval near (or before) midnight at the onset of the substorm. As the surge breaks up and moves westward, it leaves behind localized post-breakup patches that often pulsate. Figure 8.1 shows that the surge grows for about 30 minutes after onset and moves westward, disappearing at about 60 minutes after onset. In the midnight region, the patches behind the surge drift slowly eastward. The appearance of the westward traveling surge (see Figure 8.2) and the breakup of the discrete arcs result from a complex pattern of particle precipitation and broad regions of enhanced E-region densities. Observations show the intense field-aligned currents associated with the surge are carried by strong beams of precipitating electrons.

As discussed in section 5.5, another region of particle precipitation is the dayside cusp or cleft. The cusp is typically 2° to 4° wide in longitude and located at about geomagnetic latitude 78° to 80° near local noon. The dynamics of the cusp are not as clearly known as for the nighttime auroral zone. It appears that the precipitation seen in the cusp region is related more to the direct interaction of the magnetosphere and the IMF. Consistent with their magnetosheath origin, the precipitating electrons in the cusp tend to be less energetic and to vary less during disturbed periods. In most cases, cusp electrons have mean energies of only 10 eV, and rarely do any electrons have energies greater than 1 keV. As expected, such low energy electrons produce ionization in the F-region ionosphere.

Another important feature of the high-latitude ionosphere is the main electron density trough; this area of unusually low electron density forms just equatorward of the auroral oval in the nighttime ionosphere. Some researchers refer to the main trough as the sub-auroral trough. At the poleward boundary of the trough there is a sharp increase in density due to the ionizing effect of the auroral precipitation. The main trough is primarily a winter nighttime phenomena caused by two factors: (1) increased recombination due to the shorter daytime ionization periods and (2) the extended period the ions are kept in the nighttime by the high-latitude ion convection pattern. The lowest trough densities occur during weak convection conditions. In addition to having lower densities, the wintertime main trough extends over a longer local time period. At higher altitudes there is also a light ion (H^+) trough which has been observed by satellites. This trough appears to be related to the loss of ions of ionospheric origin into the magnetic tail. The light ion trough and the main trough are not always coincident.

The last major feature of the high-latitude ionosphere is the polar cap. Relatively low-energy (100 eV) electron precipitation can fill the entire polar cap; this phenomena is often called the polar rain. As a matter of comparison, the energy fluxes carried by the polar rain electrons can be as large as 10^{-3} erg/cm^2-sec; even this maximum flux is still two to three orders of magnitude less than the typical energy fluxes in the auroral oval. Observational evidence indicates the polar rain electrons travel along magnetosheath field lines and these electrons seem to enter the magnetosheath at large distances downstream from the Earth.

Satellite imagery shows that when B_z is northward Sun-aligned polar cap arcs can form intersecting both the dayside and nightside auroral oval. From space, the combination of the polar arcs and the auroral oval emissions resemble the Greek letter theta, and thus this phenomena is sometimes called the theta aurora. Observations also show that polar arcs can depart from Sun-alignment by as much as 40° and then move across the polar cap with speeds ranging from 100 to 1,000 m/sec. In addition, a four cell ionospheric convection pattern seems to be associated with the theta aurora; the sunward convection zone coincides with the position of the polar arc.

Satellite measurements confirm the inverted-V potential structures are associated with the field-aligned currents found above polar arcs. Figure 8.12 shows a possible electrodynamic configuration which can explain the formation of the polar arcs. Physically, the difference between the ion-neutral collision frequency and the electron-neutral collision frequency produces a Pedersen current flowing perpendicular to the velocity shear boundary. The convergence of the Pedersen currents at the shear reversal boundary results in an upward current flow out of the ionosphere in order to conserve charge ($\nabla \cdot \vec{J} = 0$). It appears that the upward current is carried by the field-aligned suprathermal precipitating electrons. In contrast, in cases where there is a sharp velocity differential (i.e., velocity decreases rapidly but keeps the same direction), the Pedersen current diverges at the boundary resulting in a downward current at the boundary (see problem 8-12).

Whenever B_z is southward, patches of enhanced F-region plasma originate equatorward of the dayside cusp and are then convected across the polar cap. Observations show these patches have F-region densities of approximately 10^6/cm^3 which is about 5 to 10 times larger than the ambient background density; the patches have horizontal dimensions of up to 1,000 km. Even though these patches have substantially larger densities than the normal polar cap ionosphere their emission signature is still very weak, especially in the visual wavelengths. As such, some researchers refer to the patches

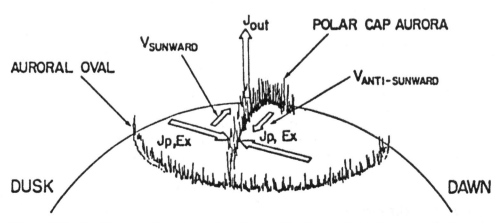

Figure 8.12 Possible electrodynamic configuration capable of generating polar cap arcs for B_z north conditions. The velocity shear ($V_{sunward}$ and $V_{anti-sunward}$) and the external electric field (E_x) drive converging Pedersen currents producing an outward flowing current (J_{out}). The precipitating suprathermal electrons forming J_{out} impact atmospheric neutrals exciting optical emissions *(after Carlson, 1990)*.

as sub-visual patches. The patches drift in the antisunward direction with speeds from 100 to 1,000 m/sec.

8.6 Sporadic E

Discussion of the sporadic E is included in a separate section because it affects all latitude regions. Sporadic E is the predominant variable feature of the quiet E-region. Typically found between 90 and 120 km, sporadic E is a transient, dense slab of ionization. It is usually 1 to 2 km thick and ranges from tens to hundreds of kilometers in diameter. Electron densities in these clouds are often two to three times that of the ambient E-layer—sufficient to noticeably alter the electron density profile at the point of occurrence. Following the cloud analogy, sporadic E may move in time. It is often found near the electrojets, the auroral oval, or in areas of intense particle precipitation. In the middle latitudes, sporadic E is often associated with meteor showers and intense thunderstorms (and squall lines). The exact connection with the latter is uncertain because of the vast difference in altitudes, but may result from electrical activity in the storm system.

Sporadic E may be thick or thin, blanketing or transparent to radio waves. The relative radio wave transparency depends not only on the electron density but also on the probe frequency and the angle of incidence. Likewise, the exact alignment (tilt and position) of the sporadic E cloud may vary, significantly altering the path of an incident radio wave. An HF radio wave may spend as much as a third of its trajectory (assumed oblique) in a thick sporadic E layer as a consequence of enhanced refraction. This changes the geometry of an oblique path by raising the virtual height of reflection (see problem 9-8).

The climatology of sporadic E is best considered by geomagnetic latitude. At low (25° or less) geomagnetic latitudes, sporadic E is primarily a daytime phenomenon showing little seasonal variation. As mentioned above, it is often found in association with the equatorial electrojet or near the geomagnetic anomalies (Southeast Asian and South Atlantic). Sporadic E is least common in the middle latitudes (25°–55° geomagnetic) and seems to be a summer daytime phenomenon. High-latitude (above 55°) sporadic E is usually found near the auroral oval/electrojet system. It results (probably) from particle precipitation and shows strong correlation with geomagnetic activity. Little seasonal variation is apparent. The most intense sporadic E (electron plasma frequencies near 8 MHz) is found in association with an intense aurora. Plasma frequencies of sporadic E associated with the diffuse aurora seldom exceed 2 to 3 MHz.

8.7 Auroral Magnetosphere— Ionosphere Coupling

Thus far, the theory we have developed attributes auroral activity to the acceleration of magnetospheric plasma into localized auroral regions by strong electric fields. These electric fields drive Birkeland currents parallel to auroral magnetic field and modify the energy spectrum of precipitating auroral electrons. We also showed that in some locations the precipitating electrons show an "inverted V" structure in a plot of the electron energy across the precipitation region where the electron energy increases to some maximum within the core of the electron precipitation region. Satellite observations also confirmed that upward-moving ion beams are aligned with the electron "inverted V" structures which implies that an electric potential drop of several to tens of kilovolts, aligned with the magnetic field, exists somewhere between the ionosphere and magnetospheric equator.

The source of these electric fields may involve one or more of the following mechanisms: (1) electrostatic double layers, (2) anomalous resistivity, and (3) magnetic mirroring of hot plasmas. Unfortunately, these mechanisms are usually treated independently and the role of ionospheric coupling is limited because of computational difficulties. We will briefly describe each mechanism and discuss the role of coupling.

Electric double layers (see Figure 5.15) are localized space charge regions which may build up large potential drops over distances on the order of several Debye lengths (~10 km for Earth). Double layers may be formed in a low density plasma when the electron temperature is different in two regions of the plasma. Instead of a smooth transition, the plasma divides itself into homogeneous regions which are separated by a double layer. At low current densities, a double layer may also form when the drift velocity equals the thermal velocity which results in an electron beam forming process which is related to the two stream instability. Theories indicate that the double layers are so thin that they are almost completely decoupled from the ionosphere.

Anomalous resistivity is an assumed region of finite resistivity along field lines (which are otherwise resistanceless) which produce the required field aligned potential drop for the Birkeland currents. Such resistance might be due to AC electric field turbulence in the auroral plasma. Many anomalous resistivity models require scale sizes of about 1 R_E for the interaction region which, in turn, requires considerable ionosphere-magnetosphere coupling.

Another possible mechanism is due to pitch angle differences between ions and electrons. If the equatorial pitch angle distributions of ions and electrons are different, then their average mirroring locations will be different, resulting in a charge separation electrostatic field. In such models ionospheric plasma becomes critically important because it partially short-circuits the very large potential drops produced by the magnetospheric plasma alone.

It is difficult to rule out any of the above mechanisms (and others) with today's observations, although the peak magnitude of the parallel electric field may be used to distinguish one model from another because model scale lengths do vary.

It is hard to believe that the ionosphere and magnetosphere are totally decoupled but the amount of coupling and its importance for the overall plasma dynamics of the near Earth space environment are still unresolved questions.

8.8 References

Aarons, J., and Basu, S. 1985. Ionospheric Radio Wave Propagation, *Handbook of Geophysics and the Space Environment*, Air Force Geophysics Laboratory, National Tech Info Service, Springfield Va, pp. 10–75.

Akasofu, S-I. 1980. Working Group Report on Geomagnetic Storms, in *Solar Terrestrial Predictions Proceedings*, R. F. Donnelly, ed. NOAA/ERL, vol. V, p. A-91.

Akasofu, S.-I. 1981. Auroral Arcs and Auroral Potential Structure, in *Physics of Auroral Arc Formation*, S.-I. Akasofu and J. R. Kan, eds., Geophysical Monograph 25, American Geophysical Union, p. 1.

Akasofu, S-I. 1982. Interaction Between a Magnetized Plasma Flow and a Strongly Magnetized Celestial Body With an Ionized Atmosphere: Energetics of the Magnetosphere. *Annual Review of Astronomy and Astrophysics*, vol. 20, p. 117.

Axford, W. I. 1981. A Review of Solar Wind-Magnetosphere-Ionosphere Coupling, in *The Physical Basis of the Ionosphere in the Solar-Terrestrial System*. AGARD Conference Proceedings No. 295, Technical Editing and Reproduction, Ltd., Harford House, London.

Baker, D. N., Akasofu, S.-I., Baumjohann, W., Bieber, J. W., Fairfield, D. H., Hones, E. W., Mauk, B., McPherron, R. L., and Moore, T. E. 1984. Substorms in the Magnetosphere, in *Solar Terrestrial Physics: Present and Future*, D. M. Butler and K. Papadopoulos, eds., NASA Ref. Pub. 1120, Chapter 8.

Basu, Su., Basu, S., MacKenzie, E., Coley, W. R., Sharber, J. R., and Hoegy, W. R. 1990. Plasma Structuring by the Gradient Drift Instability at High Latitudes and Comparison with Velocity Shear Driven Processes, *J. Geophys. Res.*, vol. 95, no. A6, p. 7799.

Bauer, S. J. 1973. *Physics of Planetary Ionospheres*. Springer-Verlag, New York.

Blanc, M. 1981. The Effects of Auroral Activity on the Midlatitude Ionosphere, in *The Physical Basis of the Ionosphere in the Solar-Terrestrial System*. AGARD Conference Proceedings No. 295, Pozzuoli, Italy 28–31 Oct 1980, Technical Editing and Reproduction, Ltd., London.

Bostrom, R. 1973. Electrodynamics of the Ionosphere, in *Cosmical Geophysics*, Egeland, Holter, and Omholt, eds. Universitetsforlaget, Oslo, Norway.

Carlson, H. C. 1990. Dynamics of the Quiet Polar Cap, *J. Geomag. Geoelectr.*, vol. 42, p. 697.

Chamberlain, J. W. 1978. *Theory of Planetary Atmospheres; An Introduction to Their Physics and Chemistry*. Academic Press, New York.

Chiu, Y. T., Anderson, R., Fennell, J., Frank, L., Hoffman, R., Hudson, M., Lyons, L., Palmadesso, P., Ungstrup, E., Vondrak, R., Williams, D., and Wolf, R. 1984. Connection Between the Magnetosphere and Ionosphere, in *Solar Terrestrial Physics: Present and Future*, D. M. Butler and K. Papadopoulos, eds., NASA Ref. Pub. 1120, Chapter 7.

Davies, K. 1965. *Ionospheric Radio Propagation*. National Bureau of Standards Monograph 80, U.S. Government Printing Office, Washington, D.C.

Gorney, D. J. 1987. U.S. Progress in Auroral Research: 1983–1986, *Rev. of Geophys.*, vol. 25, no. 3, p. 555.

Hakshmi, D. R., Aggarwal, S., and Reddy, B. M. 1980. On the Large Spatial and Local Time Electron Density Gradients in the Equatorial F-Region, in *Low Latitude Aeronomical Processes*, A. P. Mitra, ed. Pergamon Press, Oxford, p. 141.

Hanson, W. R., and Carlson, H. C. 1977. The Ionosphere, in *The Upper Atmosphere and Magnetosphere*. National Academy of Sciences, Washington, D.C.

Hardy, D. A., Burke, W. J., Gussenhover, M. S., Heineman, N., and Holeman E. 1981. DMSP/F2 Electron Observations of Equatorward Auroral Boundaries and Their Relationship with Solar Wind Velocity and North-South Component of the Interplanetary Magnetic Field. *J. Geophys. Res.*, vol. 86, p. 9961.

Hargreaves, J. K. 1979. *The Upper Atmosphere and Solar-Terrestrial Relations*. Van Nostrand Reinhold Co., New York.

Holter O., and Kildal A. 1973. Waves in Plasma, in *Cosmical Geophysics*, Egeland, Holter, and Omholt, eds. Universitetsforlaget, Oslo, Sweden.

Hunsucker, R. 1979. High-latitude E- and F-Region Ionospheric Predictions, in *Solar Terrestrial Predictions Proceedings*. U.S. Government Printing Office, Washington, D.C.

Imhof, W. L. 1986. Private communication.

Keskinen, M. J. 1984. The Structure of the High-Latitude Ionosphere and Magnetosphere, *Johns Hopkins APL Technical Digest*, vol. 5, no. 2, Johns Hopkins Applied Physics Laboratory, Laurel, Md., p. 154.

Klobuchar, J. A. 1979. Transionospheric Propagation Predictions, in *Solar Terrestrial Predictions Proceedings*, R. F. Donnelly, ed. Government Printing Office, Washington, D.C.

Knipp, D. J., Richmond, A. D., Crowley, G., DeLa Beaujardiere, O., Friis-Christensen, Evans, D. S., Foster, J. C., McCrea, I. W., Rich, F. J., and Waldock, J. A. 1989. Electrodynamic Patterns for September 19, 1984, *J. Geophys. Res.*, vol. 94, no. A12, p. 16913.

Lockwood, M., Cowley, S. W. H., Sandholt, P. E. 1990. Transient Reconnection: Search for Ionospheric Signatures, *EOS*, vol. 71, no. 20, p. 709.

McNamara, L. F. 1992. *The Ionosphere: Communications, Surveillance, and Direction Finding*, Krieger Publishing Co., Malabar, Fla.

Matsushita, S. 1959. Variations of Maximum Electron Density During Geomagnetic Storms, *J. Geophys. Res.*, vol. 64, p. 305.

Mitra, A. P., ed. 1980. *Low Latitude Aeronomical Processes*. Proceedings of Twenty-second Plenary Mtg. of COSPAR, Pergamon Press, Oxford.

Moyer, V. 1976. *Physical Meteorology*. Texas A & M University, College Station, Tex.

Neske, E., Rawer, K., and Rebstock. 1980. Longitudinal Variation of Peak Electron Density at Low Latitudes, in *Low Latitude Aeronomical Processes*, A. P. Mitra, ed. Pergamon Press, Oxford, p. 107.

Nisbet, J. S. 1975. Models of the Ionosphere, in *Atmospheres of Earth and the Planets*, B. M. McCormac ed. D. Reidel Pub. Co., Boston.

Ossakow, S. L., Burke, W., Carlson, H. C., Gary, P., Heelis, R.,. Keskinen, M., Maynard, N., Meng, C., Szuszczewicz, E., and Vickery, J. 1984. High Latitude Ionospheric Structure, in *Solar Terrestrial Physics: Present and Future*, D. M. Butler and K. Papadopoulos, eds., NASA Ref. Pub. 1120, Chapter 12.

Rawer, K. 1952. Calculation of Sky-wave Field Strength. *Wireless Engineer*, vol. 29, p. 287.

Rawer, K. Harnischmacher, E., and Eyfrig, R. 1981. The Day-by-Day Variability of the Ionospheric Peak Density, in *The Physical Basis of the Ionosphere in the Solar-Terrestrial System*, AGARD Proceeding No 295.

Robinson, R. M. and Vondrak, R. R. 1985. Characteristics and Sources of Ionization in the Continuous Aurora, *Radio Science*, vol. 20, no. 3, p. 447.

Rottger, J. 1981. The Seasonal and Geographical Variation of Equatorial Spread—F Irregularities Influenced by Atmospheric Gravity Waves and Electric Fields Due to Thunderstorms, in *The Physical Basis of the Ionosphere in the Solar-Terrestrial System*, AGARD Proceeding No. 295.

Rush, C. 1979. Report of the Mid- and Low-Latitude E and F Region Working Group, in *Solar Terrestrial Predictions Proceedings*. U. S. Government Printing Office, Washington, D.C.

Schunk, R., Barakat, A. R., Carlson, H. C., Evans, J. B., Foster, J., Greenwald, R., Kelley, M. C., Potemera, T., Rees, M. H., Richmond, A.

D., and Roble, R. G. 1984. Assessment of Plasma Transport and Convection at High Latitudes, in *Solar Terrestrial Physics: Present and Future*, eds D. M. Butler and K. Papadopoulos, NASA Ref. Pub. 1120, Chapter 11.

Singh, M., and Gurm, H. S. 1980. Day to Day Variability in the Low-Latitude Ionosphere, in *Low Latitude Aeronomical Processes*, A. P. Mitra, ed. Pergamon Press, Oxford, p. 149.

Sojka, J. J. 1989. Global Scale, Physical Models of the F Region Ionosphere, *Rev of Geophys*, vol. 27, no. 3, p. 371.

Sojka, J. J., Raitt, W. J., and Schunk, R. W. 1981. Plasma Density Features Associated with Strong Convection in the Winter High-Latitude F-Region, *J. Geophys. Res.*, vol. 86, p. 6908.

Szuszczewicz, E. P. 1986. Theoretical and Experimental Aspects of Ionospheric Structure: A Global Perspective on Dynamics and Irregularities, *Radio Science*, vol. 21, no. 3, p. 351.

Tascione, T. F., Flattery, T. W., Patterson, V. G., Secan, J. A., and Taylor, J. W. 1979. Ionospheric Modeling at Air Force Global Weather Central, in *Solar Terrestrial Predictions Proceedings*. U. S. Department of Commerce, U. S. Government Printing Office, Washington, D.C.

Thrane, E. V. 1973. Radio Wave Propagation, in *Cosmical Geophysics*, Egeland, Holter, and Omholt, ed. Universitetsforlaget, Oslo, Sweden.

Thrane, E. V. 1979. D-Region Predictions, in *Solar Terrestrial Predictions Proceedings*. U.S. Government Printing Office, Washington, D.C.

Townsend, R. E. 1982. *Source Book of the Solar-Geophysical Environment*. AFGWC/WSE, Offutt AFB, Neb.

Valladares, C. E. and Carlson, H. C. 1991. The Electrodynamic, Thermal, and Energetic Character of Intense Sun-Aligned Arcs in the Polar Cap, *J. Geophys. Res.*, vol. 96, no. A2, p. 1379.

Vondrak, R. R. 1979. Magnetosphere-Ionosphere Interactions, in *Solar Terrestrial Predictions Proceedings*. U.S. Government Printing Office, Washington, D.C.

Weber, E. J., Klobuchar, J. A., Buchau, J., Carlson, H. C., Livingston, R. C., DeLa Beaujardiere, O., McCready, M., Moore, J. G., and Bishop, G. J. 1986. Polar Cap F Layer Patches: Structure and Dynamics, *J. Geophys. Res.*, vol. 91, no. A11, p. 12121.

Whalen, J. A., O'Neil, R. R., and Picard, R. H. 1985. The Aurora, in *Handbook of Geophysics and the Space Environment*, ed. Adolph S. Jursa, Air Force Geophysics Laboratory, National Technical Information Service, Springfield, Va., Chapter 12.

Wright, J. W. 1962. Dependence of the Ionospheric F Region on the Solar Cycle. *Nature*, vol. 194, 461.

8.9 Problems

8-1. The text states that polar cap F-region patches convect across the polar cap at speeds ranging from 100 to 1,000 m/sec. Estimate the magnitude and direction of the polar cap electric field needed to drive this convection. Is this electric field consistent with the polar cap potential drop known to exist for B_z south conditions? Explain your answer.

8-2. Distinguish between regular ionospheric variation versus "non-normal" variations. Include informa-

tion on geographic location, intensity, duration, seasonal behavior, and physical processes.

8-3. Describe the time sequence of an ionospheric storm at various geomagnetic latitudes. Do ionospheric storms differ from the magnetospheric substorm discussed in Chapter 5?

8-4. What is the equatorial anomaly and how is it related to the "fountain effect"? Is the equatorial fountain effect different from the cleft fountain described in section 5.5? Explain.

8-5. Will the Southeast Asia and/or South Atlantic anomalies have any serious consequences for low-altitude satellite operations? How about for communications?

8-6. Differentiate between equatorial spread F and high-latitude spread F. Include information on source mechanisms, duration, geographic extent, and seasonal dependence.

8-7. What is the midlatitude winter anomaly and what are the physical processes controlling this phenomena? Does the anomaly have any solar cycle dependence?

8-8. What is the auroral oval and why is it so important for understanding magnetospheric-ionospheric coupling? What is the sub-auroral trough and why does it exist?

8-9. Compare and contrast the following field-aligned electric field mechanisms: 1) electrostatic double layers, 2) anomalous resistivity, and 3) magnetic mirroring.

8-10. Describe the important features and characteristics of sporadic E. Include information on sources, intensity, location, and duration.

8-11. (a) Starting with equation 1.31 derive equation 8.1 for the equatorial F-region ion drift velocity for the conditions in Figure 8.5. (b) Derive the analogous equation for the electrons. (c) Explain why the accumulation of charge shown in Figure 8.5(b) occurs.

8-12. In section 8.5, the text states that the polar cap Pedersen currents converge at velocity shear boundaries and diverge at velocity differential boundaries. Derive the necessary equations showing this statement is true. In your analysis have you made any assumptions about the formation or loss of ionospheric ions and electrons? Does the "Pedersen current mechanism" work anywhere else in the ionosphere? Explain your answer.

Chapter 9

Radio Wave Propagation in the Ionosphere

9.0 Introduction

The ionosphere can affect the transmission of radio waves in at least two ways. First, if conditions are suitable, the charged particles can remove energy from an electromagnetic wave and thus attenuate the signal. In the limiting case, the energy of the wave can be absorbed completely. Second, a wave traveling a path along which the electron density is not constant undergoes a change in the direction of its propagation. In certain circumstances, the radio wave can be reflected. We will examine each of these effects.

9.1 Attenuation of Radio-Frequency Waves

When a radio-frequency (RF) wave enters the ionosphere and encounters a significant concentration of free electrons, some of the energy of the wave is transferred to the electrons, which are then set into oscillation at the frequency of the RF wave. In turn, the electrons can lose some of this energy by colliding with neutral atmospheric particles. However, if there are no such collisions, the oscillating electrons will reradiate electromagnetic energy at the same frequency, thus restoring most of the energy of the RF signal.

When the atmospheric density is appreciable, say about 10^{-4} of the sea-level value, as it is between 50 and 65 km, the atmosphere contains such a relatively large concentration of neutral particles that collisions with free electrons will occur at a significant rate. In these collisions, much of the energy of oscillation that the electrons have acquired from the radio wave is transferred to the neutral particles and appears as heat. Consequently, the RF signal is attentuated: the transmitted signal is weakened. If the rate of collison between electrons and neutral particles is large enough, essentially all of the energy of the RF wave is transferred to the neutral atmosphere, and the signal is then said to be absorbed completely.

Appreciable loss of signal strength occurs in the atmosphere when there are a sufficient number of electrons *and* neutral particles, the former to take up the energy from the electromagnetic wave, the latter to remove the energy from the oscillating electrons via collisions. This is the situation that prevails in the D-region. The density of neutral particles is about 10^{-4} to 10^{-5} of that of sea level, and the electron density, while not as large as in the E- or F-regions, is enough to allow considerable interaction with RF waves. Therefore, appreciable attenuation of radio signals can occur in the D-region. Other conditions being equal, the lower frequency (longer wavelength) waves are attenuated more than higher frequency (shorter wavelength) waves. In the E- and F-regions, the atmospheric density is much less than in the D-region, and, although the electron density is larger, the number of collisions is smaller. As a general rule, radio waves suffer little attenuation of signal strength due to collisions when passing through the higher levels of the ionosphere. Most of the collisional attenuation that does occur takes place at the lower altitudes, i.e., in the D-region.

This discussion has been based on the interaction between electromagnetic waves and electrons. Of course, both positive and negative ions, in principle, also can be set into oscillation by RF waves and so can take up energy from the latter. This energy then may be transferred by collision to neutral particles. However, because of their much greater masses (the minimum of which, in the case of the proton, is some 1840 times as great as that of the electron), the ions do not oscillate as readily as do the electrons and, thus, are far less effective in removing energy from the RF waves. Consequently, the attenuation due to ions in the ionosphere can be neglected for most purposes.

9.2 Reflection of Radio-Frequency Waves

The second effect of the ionosphere on radio waves, namely the change of direction of propagation, arises from the effect free electrons have on the velocity of an electromagnetic wave. For RF waves traveling in the *vertical* direction through a nonmagnetic, ionized medium, the *phase* velocity u (cm/sec) is related to the velocity of light in vacuum c by

$$u = c\left[\sqrt{1 - \frac{N_e q^2}{\pi m \nu^2}}\right]^{-1} \quad (9.1)$$

where N_e is the electron density in cm^{-3}, q is the magnitude of the charge on the electron, m is the mass of the electron in grams, and ν is the frequency of the RF wave. Thus for ν = constant, the phase velocity increases with an increase in electron density. Remember, that wave information travels at the group velocity, not the phase velocity, and only the group velocity is limited by the speed of light. When we insert the constant values of q, m, and π and express the frequency in megahertz, 9.1 becomes

$$u = c\left[\sqrt{1 - \frac{8.1 \times 10^{-5} N_e}{\nu^2}}\right]^{-1}. \qquad (9.2)$$

The refractive index n of the ionized medium for RF waves equals c/u, hence,

$$n = \sqrt{1 - (N_e q^2/\pi m \nu^2)} \qquad (9.3)$$

or

$$n = \sqrt{1 - (8.1 \times 10^{-5} N_e/\nu^2)}. \qquad (9.4)$$

It is evident from equation 9.3 that, for waves of a given frequency, the refractive index decreases in the case of a wave passing from a region of lower electron density to one of higher electron density. The greater the N_e, the smaller the value of the radical, and, thus, the smaller the refractive index. Therefore, Snell's law tells us that a *beam* or *ray* of a radio signal will be refracted in the manner illustrated in Figure 9.1; the angle α' becomes larger than the angle α as the ray propagates along a path in which the electron density increases—as it does in going from the D- to E- and from the E- to the F-regions, respectively. The same conclusion can be drawn from the fact that the phase velocity increases with electron density, so that the upper part of an oblique propagating wave travels faster than the lower part, and thus the wave is bent away from the vertical direction ("downward").

For a given electron density, a high-frequency (short wavelength) wave will suffer less refraction than one of lower frequency (longer wavelength) because n is closer to one in the high frequency case. Equations 9.1 and 9.3 were derived on the simplifying assumption that the electrons undergo no collisions and that there is no magnetic field present, but these restrictions will have no appreciable effect on the

conclusions to be drawn from these relationships (see Problem 9-10).

Now consider the case in which a radio wave of constant frequency enters the ionosphere from below. At first, the electron density is small and the wave travels essentially with the speed of light in vacuum. But as the concentration of electrons increases with altitude, and as long as the RF frequency remains constant, the phase velocity must increase while the refractive index decreases. Eventually, the quantity under the radical in equation 9.4 can go to zero when

$$N_{ec} = \frac{\nu^2}{8.1 \times 10^{-5}} = 1.2 \times 10^4 \, \nu^2. \qquad (9.5)$$

At this value of the electron density, represented by N_{ec} and called the critical electron density, the RF wave no longer can be propagated in the forward direction (upward), and it is reflected back to Earth. The critical frequency is the highest frequency that can be reflected by a given electron density:

$$\nu_c = 9 \times 10^{-3} \sqrt{N_{ec}}. \qquad (9.6)$$

In the D-region, for example, the maximum observed value of electron density is about 10^3 cm^{-3}, and thus, the highest frequency that can be reflected (for vertical propagation) is about 0.28 MHz (or 280 kHz). On the other hand, the maximum electron density of the F-region is roughly 2×10^6 cm^{-3}, so that vertically propagating radio waves with frequencies up to 13 MHz are reflected. Therefore, waves of sufficiently high frequency will pass through the D-region but may be reflected in higher regions where the electron concentrations are greater. For radio waves transmitted in a vertical or near-vertical direction, equations 9.5 and 9.6 give the appropriate critical electron density and critical frequency, respectively. However, for purposes of long-range communication, the radio signal must have a relatively large angle of incidence α with respect to the vertical. In that case, the above equations require modification.

In passing through the ionosphere, where the electron density increases gradually with height, the radio wave is not reflected specularly (as in a mirror) but is refracted continuously, and the actual path resembles that shown in Figure 9.2. for oblique propagation, the refractive index is decreased to zero at a lower electron density than would be the case for vertical incidence. Under these circumstances, equation 9.5 takes the form

$$N_{ec} = 1.2 \times 10^4 \, \nu^2 \cos^2 \alpha. \qquad (9.7)$$

The corresponding form of equation 9.6 is

$$\nu_c = \frac{9 \times 10^{-3}}{\cos \alpha} \sqrt{N_{ec}} \qquad (9.8)$$

so that the critical frequency is increased when the angle of incidence is greater than 0°. For example, at $\alpha = 60°$, the critical frequency for reflection by the F-region is increased

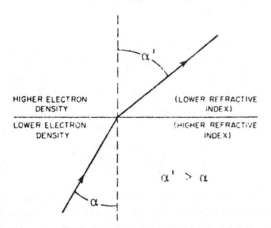

Figure 9.1 Refraction of radio waves by the ionosphere due to changes in the electron density.

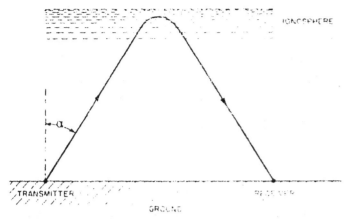

Figure 9.2 Path of "reflected" radio wave resulting from ionospheric refraction.

from 13 to 26 MHz. The critical frequency is also known as the plasma frequency of the critical level.

Unless it is absorbed, a radio wave of a given frequency transmitted upward will continue to travel through the ionosphere until it reaches its critical level where the electron density has the value given by equation 9.7. Then the wave will be turned back. The higher the frequency, the greater the electron density at which reflection will occur. As seen in an earlier chapter, the electron density increases more or less continuously up to an altitude of roughly 300 km, after which it decreases with height. Consequently, the higher the frequency of the radio waves, the farther the distance they can travel into the ionosphere, up to a height of 300 km or so, before undergoing reflection. The limiting frequencies for

which reflection is possible at any particular level are given in equation 9.8. Frequencies above this limiting value will pass completely through the ionosphere and out into space (see Figure 9.3).

Transmission of radio signals from above the atmosphere, as in the telemetry of data from artificial satellites, requires merely a frequency high enough to penetrate the F2-peak. The problem is quite similar to that of upward transmission, since the electron density increases downward from the upper ionosphere to the peak value near 300 km and then decreases.

Next we will include the effects of a magnetic field on the propagation equations. If we propagate the wave parallel to a background magnetic field, equation 9.3 becomes

$$n = \sqrt{1 - \frac{\omega_0^2}{\omega^2 \pm \omega\omega_g}} \qquad (9.9)$$

where we have used the notation

$$\omega_0 = \text{plasma frequency} = \sqrt{\frac{4\pi N_e e^2}{m}}$$

$$\omega = \text{radio frequency} = 2\pi\nu$$

$$\omega_g = \text{gyrofrequency} = \frac{eB}{mc}$$

Again we are using cgs units for convenience. The magnetic field strength B used in the gyrofrequency is measured along the direction of wave propagation. The ± sign in equation 9.9 means that a radio wave probing the ionosphere with a magnetic field splits into two separate propagation modes (analogous to an optical wave in a birefringent medium).

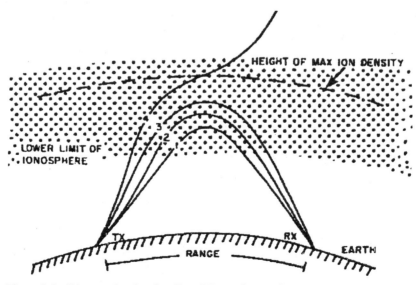

Figure 9.3 Diagram showing the effect different frequencies have on several ray paths for a fixed distance between transmitter (TX) and receiver (RX). As the frequency increases, the ray penetrates deeper into the ionosphere until it eventually escapes *(after McNamara, 1985)*.

Therefore, we say that the Earth's ionosphere has magnetic birefringence. The index of refraction for the so-called L wave is given by

$$n_L = \sqrt{1 - \frac{\omega_o^2}{\omega^2 + \omega\omega_g}} \qquad (9.10)$$

which gives the critical reflection frequency ($n_L = 0$) as

$$\omega_L = \left(\omega_o^2 + \frac{\omega_g^2}{4}\right)^{1/2} - \frac{\omega_g}{2}. \qquad (9.11)$$

The R wave is given by

$$n_R = \sqrt{1 - \frac{\omega_o^2}{\omega^2 - \omega\omega_g}} \qquad (9.12)$$

and the critical reflection frequency becomes

$$\omega_R = \frac{\omega_g}{2} + \sqrt{\frac{\omega_g^2}{4} + \omega_0^2}. \qquad (9.13)$$

A linearly polarized radio wave transmitted into the ionosphere parallel to the magnetic field becomes split into two components: right and left circularly polarized, which propagate at different velocities and are reflected at different heights. The electric field vector for the right, or R, wave rotates clockwise in time as viewed along the direction of the magnetic field. Conversely, the electric field vector for the left, or L, wave rotates counterclockwise.

If instead, the radio wave propagates perpendicular to the background magnetic field B, with the wave electric field parallel to B, then the index of refraction for the ordinary wave simply becomes equation 9.3, or in terms of the plasma frequency, can be written as

$$n = \sqrt{1 - \frac{\omega_0^2}{\omega^2}}. \qquad (9.14)$$

If the wave electric field is perpendicular to B, then the index of refraction for the extraordinary wave becomes

$$n = \sqrt{1 - \frac{\omega_0^2(\omega^2 - \omega_0^2)}{\omega^2(\omega^2 - \omega_h^2)}} \qquad (9.15)$$

where ω_h is called the upper hybrid frequency and is given by

$$\omega_h = \sqrt{\omega_0^2 + \omega_g^2}. \qquad (9.16)$$

9.3   Determination of Electron Densities

The usual procedure for determining electron-density profiles from ground stations is based on radio-echo (or time-delay) techniques. The instrument that is used is the vertical incidence ionospheric sounder or ionosonde. A radio pulse of known frequency is transmitted vertically and the elapsed time of the echo is indicated by an oscillograph that is coupled to a receiver of the same frequency. Since the pulse rate is very high, and the frequency of transmission is varied rapidly from about 1 to 20 megahertz, this sweep-frequency technique results in a plot called an ionogram (see Figure

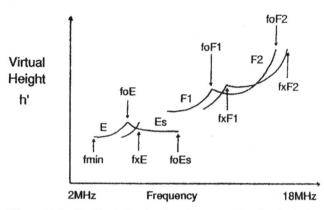

Figure 9.4   Idealized sketch of an ionogram showing both the ordinary and extraordinary wave traces; also shown is an example of a sporadic E trace (Es) and the critical frequency of the sporadic E layer (f_oEs) *(after Townsend, 1982)*.

9.4). The right side of the figure would show the virtual heights of each layer [e.g. the E layer (h'E), F1 layer (h'F1), and F2 layer (h'F2)]. Across the bottom of the figure would be the critical frequencies for each layer. For example, foF2 is the critical frequency of the F2 layer determined from the propagation properties of the ordinary wave and fxF2 is the critical frequency associated with the extraordinary wave.

If the signal traveled all the way with the speed of light, as it does to a close approximation in the sparsely ionized region below the ionosphere, the delay time would be a direct measure of the height at which reflection occurred for each frequency. But since radio pulses are slowed down in the heavily ionized regions, the average speed is less than that of light in a vacuum. (The pulses travel with the group velocity, which decreases with increases in electron density.) Hence, the real height of reflection is less than the virtual height that is derived directly from the observed delay time (see Figure 9.5). From the frequency of the transmitted (and reflected) pulse, the electron density at which reflection occurs is determined from equation 9.8 and the corresponding virtual height can be corrected to yield the true height for that density. The virtual height is sometimes called the equivalent height.

The chief drawback to ground-based sounding for investigating the ionosphere is the inability to probe beyond the altitude of maximum electron density in any region. This limitation has not been serious for the D-, E-, and F1-regions, but it hampers investigation of the F2-region above the level of peak electron density. This is of small concern from the standpoint of radio transmission, but a knowledge of electron densities at greater altitudes is important in connection with studies of the atmosphere.

A modified, ground-based sounding technique using high powered radars permits probing the ionosphere both below and above the F2 maximum. When a strong, short pulse of RF energy is transmitted upward from the ground, the

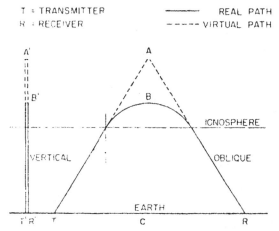

T = TRANSMITTER ———— REAL PATH
R = RECEIVER ----- VIRTUAL PATH

Figure 9.5 Virtual height. The figure on the left is for vertical propagation (both the transmitter T' and receiver R' are colocated). Using the time between signal transmission and reception, with the simplified assumption that radio waves travel at the speed of light, one would say that the signal reflected from the ionosphere at height A' (virtual height). In fact, the real reflection height is lower because the radio wave slows down as it encounters the increasing electron density at higher altitudes. The figure to the right is for oblique propagation, and the virtual height A also represents the effective control point.

electromagnetic waves undergo weak incoherent scattering from individual electrons in the ionosphere. In incoherent scattering, each scattering particle acts independently of the others, and the scattered waves do not interfere with or enhance each other. This results in a small amount of radio energy being scattered back to the ground. The weak signal is called an incoherent backscatter echo and the transmitter/ receiver is usually referred to as an *incoherent backscatter radar*. The echo signal strength is proportional to the electron density and the delay time is a measure of the height at which the scattering occurred. Radar waves of frequency well over 30 megahertz can penetrate the F2-region and yield information above the F2 peak.

9.4 Bands and Modes

Radio waves are subdivided into bands according to their frequency or wavelength. Commonly used terminology is summarized in Table 9.1. Microwave is a general term applying all frequencies above about 200 MHz.

9.5 Propagation Modes

Because of their varying interactions with the ionosphere, radio waves may propagate in one or more different manners, or modes. Moreover, a given frequency may propagate by several modes simultaneously. Which modes predominate are determined by the ionosphere and the transmitter and receiver. Figure 9.6 provides examples of the primary propagation modes for radio waves. These include the surface wave, ground-reflected wave, direct wave, tropospheric wave, and sky wave.

The surface wave is the component which propagates along the atmosphere/earth junction. The efficiency of this mode of propagation depends on the conductivity of the surface over which it is passing. Water is a relatively good conductor, and this mode is important for overwater propagation when both the transmitting and receiving antennas are close to the surface. Some frequency bands will penetrate the surface (greatest penetration occurs for the lowest frequencies) and become rapidly attenuated. Whereas, other frequencies are reflected in varying degrees at this boundary. Surface wave range depends on both surface conductivity and radio wave frequency.

Generally, neither the ground-reflected wave nor the direct wave is effective over significant distances. The ground-reflected wave has an effective range, depending on antenna height, of a few miles. In fact, for some bands it may create a significant source of interference. Surface and ground-reflected wave modes are most important for the long wavelength bands (ELF, VLF, and LF).

By comparison, the direct (or line-of-sight) wave is probably the most reliable mode of radio wave propagation. Curvature of the Earth effectively limits its range to a few

Table 9.1 Radio wave bands

Frequency Range	Wavelength Range		Common Name
20 Hz–3 kHz	Greater than 100 km	(ELF)	Extremely Low Frequency
3 kHz–30 kHz	100 km–10 km	(VLF)	Very Low Frequency
30 kHz–300 kHz	10 km–1 km	(LF)	Low Frequency/Long Wave
300 kHz–3 MHz	1 km–100 m	(MF)	Medium Frequency/Medium Wave
3 MHz–30 MHz	100 m–10 m	(HF)	High Frequency/Short Wave
30 MHz–300 MHz	10 m–1 m	(VHF)	Very High Frequency
300 MHz–3 GHz	1 m–10 cm	(UHF)	Ultra High Frequency
3 GHz–30 GHz	10 cm–1 cm	(SHF)	Super High Frequency
30 GHz–300 GHz	1 cm–1 mm	(EHF)	Extremely High Frequency

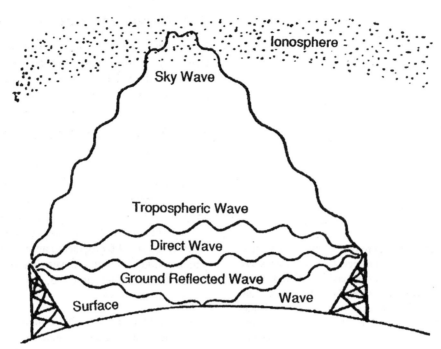

Figure 9.6 Various types of propagation modes available for radio waves. The exact mode of propagation will depend on signal frequency and atmospheric conditions (in particular, ionospheric conditions) *(after Prochaska, 1980)*.

tens of miles. Direct wave propagation is most effective for the medium frequency bands and above (MF, HF, VHF, UHF, etc.).

The tropospheric wave is, at first glance, apparently similar to the sky wave. Its refraction in the atmosphere usually occurs well below the ionosphere and is unrelated to electron densities. Steep vertical temperature or humidity gradients (often inversions) are the primary refracting causes. This propagation mode produces the effect called radar ducting, which can be especially important for UHF bands and higher frequencies. It is most effective for higher frequencies, but even here, range is limited. Tropospheric scattering can be a nuisance when path geometry is important (as in radar systems). In other situations, it is a primary mode of propagation.

Long-range propagation (in excess of about 1500 km) without intermediate repeater stations must use sky wave. For HF frequencies, sky wave propagation relies on the refractive properties of the ionospheric plasma to bend the radio wave back to Earth beyond the Earth's curvature. The range and path geometry of sky wave propagation depends on the transmitter takeoff angle (complement of the angle of incidence, α) and a number of ionospheric parameters. The sky wave is refracted by the ionosphere at the path control point. As it moves downward, the signal will be reflected by the Earth's surface, with seawater providing a stronger reflection and forested areas tending to absorb more radio energy. This "bounce" permits a range extension beyond 4000 km by means of a multihop path (see Figure 9.7). Each bounce/

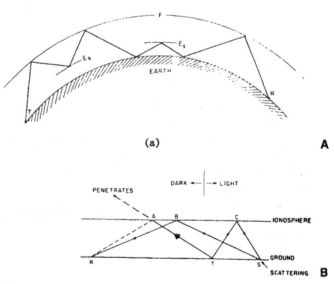

Figure 9.7 Multihop paths. (A) Propagation path that is affected by sporadic E (E_s). Notice that in one case the signal never reaches the ground and in the other E_s reflection, the signal makes a short hop. (B) Propagation from a transmitter T at the terminator (dark/light boundary) to a receiver at R. The direct path T to A passes through the weak nighttime ionosphere (i.e., propagation frequency too high). However, a scattered signal from path TC returns across the terminator and strikes the nighttime ionosphere with a larger angle of incidence so that the signal is reflected down to R. In such an example, the communicators would not know the true signal path, and they would probably believe the signal is going directly from the transmitter to the receiver via a single hop nighttime path *(after Townsend, 1982)*.

refraction will scatter the incident radio energy; some will be sent forward, in the direction of propagation, and some will be reflected toward the transmitter. A small, but non-zero fraction will be scattered to the left and right of the desired path. Finally, some of the energy will be absorbed in the surface and the ionosphere, and some will escape out of the Earth's atmosphere. Most radio energy will propagate along a great-circle path away from the transmitter in a direction dependent on the antenna azimuth. However, ionospheric irregularities (tilts, or horizontal gradient in the ionospheric electron density) may produce non-great-circle propagation modes. The more hops a path contains, the greater the potential for path irregularities.

Path irregularities usually occur in regions of strong horizontal (as opposed to the normal, vertical) electron density changes. Such variations are common near the auroral oval/subauroral trough and in the subequatorial regions. In these regions, strong horizontal gradients persist over long periods of time. Furthermore, short-lived discontinuities are thought to appear and disappear throughout the ionosphere. Many probably last no more than 1–2 hours, and are often associated with the sunrise and sunset terminators. To confuse the results even further, a portion of the wave energy may maintain its original ground track, while the other portion is deflected in another direction. These unintentional path changes can be used to the operator's benefit if he is aware of their existence and structure.

9.6 Radio Wave Transmission and Noise

Radio waves are used for many different purposes, but the underlying purpose is to convey information from one point to another without establishing a wire link between the points. The information may be music, a television picture, an encrypted set of numbers, or just a series of dots and dashes. A transmitter normally operates at a given frequency, called the carrier frequency, and this would be the nominal frequency at which the station is tuned in.

Unfortunately, very little information could be transmitted at a single frequency. Therefore, a channel consists of a small band of frequencies on either side of the carrier frequency. These bands are termed the upper and lower sidebands and the total bandwidth of the channel is the sum of the frequency range of the two sidebands. The finite bandwidth of each radio wave channel automatically limits the number of users of a given frequency band. If more than one user is assigned the same channel, the potential for interference will exist; the degree of interference depends on the range of the frequency band in use and the proximity of the transmitters.

While bandwidth limits the amount of information which can be transmitted by radio wave, the signal-to-noise ratio (S/N) determines what will be received. The larger this ratio, the easier it is to differentiate between signal and background noise. Radio wave systems require a minimum S/N in order

to operate successfully and S/N can be increased by increasing signal strength, reducing noise, or both.

Signal strength is initially dependent on transmitter power and antenna efficiency (both receiver and transmitter). Antenna efficiency can be increased by using a directional antenna which can concentrate available power into the desired direction. Selection of antenna size is also an important part of efficiency. For example, at least one element of the antenna should, ideally, be equal in length to a full or half wavelength of the radio wave to be used.

Radio noise, both manmade and environmental, is the second variable in our S/N ratio. If radio noise could be eliminated, infinite receiver amplification would be possible and, therefore, infinitesimal transmitter power would be required. However in practice, receiver amplification increases noise and thus there are limits to its use. Manmade noise sources can at least be identified. Most are associated with interference, electric motors, or electronic noise internal to the radio itself. The latter type is often the dominant noise source on frequencies above about 300 MHz. Narrow bandwidth, directional antennas, and high power are the only real solutions to manmade noise other than eliminating it. Unfortunately environmental noise, such as lightning discharges, cannot be eliminated.

9.7 Fading and Absorption

The environment can alter the signal and noise levels in other ways as well, and sky waves are by far the most vulnerable. The conductivity of the Earth's surface near the ground "bounce" point and the ionospheric conductivity near the path control point both affect the amount of wave energy forwarded to the receiver. Likewise, absorption and fading can reduce the received signal and noise levels. These effects are generally frequency dependent. Fading usually results from path variations, whereas absorption is generally a consequence of conversion of radio energy into heat in the ionosphere. Absorption reduces both signal and environmental noise. Fading may affect both, but is more often a periodic change in signal strength alone.

Fading commonly results from changes in path geometry, but other origins are also possible. Interference fading results when a signal from a given transmitter is made to follow two or more separate paths to the receiver. Horizontal electron density gradients (ionospheric tilts) may cause this by splitting the radio wave energy into several signals. Since the separate paths are often of different lengths, the signals will arrive slightly out of phase with one another. Unless electronic sophistication is employed, the receiver unknowingly accepts both signals and the resulting signal strength may vary due to destructive and constructive interference. This type of fading often has a period of seconds and is commonly known as multipath. Polarization fading is similar in period and origin, and results from slight changes in the signal's polarization which occur during the reflection by the ionosphere. Again, when the receiver combines the signals, fading results. Both polarization and interference fading occur near high latitudes because ionospheric tilts are common here. Unlike fading, absorption typically reduces both the desired and unwanted signal levels at the receiver.

9.8 The Long Waves—ELF, VLF, LF

D-layer plasma frequencies effectively prevent the entry of radio frequencies below about 300 kHz into the ionosphere. In short, these frequency bands are trapped between the Earth's surface and the ionosphere. Sky wave propagation at these frequencies is by a pseudo waveguide mode between ground and ionosphere, and considerable range and good signal strengths result. This waveguide also insulates these bands from much of the ionosphere's variability, and ELF through LF often remain usable even during an ionospheric storm.

Unfortunately, these advantages are offset by several serious disadvantages. Probably the most serious of these is the limited bandwidth available. Noise levels are high in these bands, so high power levels are required. This is particularly critical for the longest waves where equipment is not only expensive, but also very large. Some ELF systems require antennas hundreds of meters long, several megawatts of power, and employ amplifiers which are as large as a truck.

The high cost and sometimes ungainly size of these systems limit their use to certain specialized applications. Long-range navigation and subsurface (ocean) communication are their primary uses. These systems provide particularly good range over a good conductor, such as water. Moreover, signal penetration into a conductor is inversely proportional to wavelength. VLF will penetrate 10 to 15 m into seawater, and ELF can theoretically be received at depths in excess of 100 m. While a submarine might not have the space for an ELF transmitter, an ELF receiver and trailing wire antenna would permit reception of signals while remaining safely submerged. Use of LF or higher frequencies would require that at least the antenna mast be above the water's surface. VLF serves surface navigation through such systems as loran and OMEGA, where D-layer stability is essential to accurate positioning of the receiver.

The great overwater range permits a few VLF stations to provide nearly global navigation capability. LF experiences markedly higher absorption and is used for short-range direction finding, particularly by aircraft near airports. This permits lower power LF systems to function with acceptable accuracy.

9.9 Medium Frequencies—MF

The 300 kHz-3 MHz frequency band is the lowest band capable of at least partial penetration into the ionosphere. A comparison of plasma frequencies reveals that MF should pass through the normal D-layer and be reflected by or refracted in the E-layer. However, since MF frequencies do not significantly exceed the critical frequency of the D-layer, MF signals experience considerable collisional absorption during daylight hours. In fact, surface wave and line-of-sight are the primary means of daytime propagation for an MF system. Long distance (beyond a few hundred miles) communication is impossible for MF systems during the daylight unless transmitter power is very high. At night, the situation changes because recombinations eliminate much of the D-region. With the resulting drop in absorption, even low-power circuits can communicate successfully over great distances.

Long distance communication is not without its problems. The MF band contains nondirectional airport radio beacons at its low end and the commercial (AM) broadcast band at its upper end. Considerable frequency congestion exists in the MF band, and a number of stations are assigned closely spaced or overlapping channels. These stations are widely separated in geographical location, so there is little interference during daylight hours. At night, a very different situation develops. AM broadcast stations can often be received over considerable distances and distant stations can thus interfere with local stations. The result, at least in the United States, is a complicated allocation of (1) assigned frequency (channel), (2) authorized power, and (3) limited broadcast times.

Certain stations (AM) in the United States are assigned clear channels. These stations are authorized to operate at high power (perhaps 50 kW) 24 hours a day and are assured that no other stations are authorized to use their frequency within the United States. Consequently, clear channel stations are unlikely to interfere with local stations (or be interfered with by them) and may be received over a considerable distance—particularly at night.

Other U.S. stations are not similarly assured of interference-free operation. In fact, in the U.S. widely separated stations are often assigned the same frequency. In order to minimize interference with stations operating on the same frequency (actually a channel, since a finite bandwidth is assigned), several restrictions or combinations of restrictions are employed. These stations are usually limited in output power. Moreover, they are usually required to operate at even lower power (perhaps only 500 watts) during nighttime hours or go off the air entirely between sunset and sunrise. The latter restriction is often employed with new stations or stations located very near previously existing stations. Even these restrictions don't always ensure interference-free reception due to lack of receiver selectivity or transmitter frequency drift. The Federal Communications Commission (FCC) closely monitors the carrier frequency and effective bandwidth of commercial broadcast stations. The move toward stereo broadcasts on AM stations will increase the already acute frequency congestion in the MF band unless it is accompanied by increased technology. (Remember: More information must be transmitted to generate stereo than for monophonic sound; so greater bandwidth is required.)

Disturbances of the ionosphere also affect MF operations. Yet, these effects are certainly limited by comparison with the impact of solar geophysical activity on the neighboring LF and HF bands. The structure of the ionosphere itself explains this "insulation" of MF frequencies. Since MF does not penetrate the D-region, it is not particularly sensitive to D-layer height variations. The increase in D-region electron density associated with a solar flare does increase absorption of the MF signals, but this absorption is really noticeable only on frequencies near the top end of the MF band operating with considerable power. Lower frequency circuits are normally completely absorbed in the D-region even in the absence of a solar flare, so increased flare emission is of little concern to most MF systems. Occasionally the D-region enhancement due to a flare will be sufficient to permit the layer to reflect MF signals. For a brief period of time, the affected MF circuits will experience a significant range and signal strength increase due to their exclusion from the D-region. The result will often be increased interference by distant stations also using the affected frequencies.

Interference may also result from the radio bursts associated with solar flares. Many solar radio noise storms are concentrated in the upper LF and lower MF bands, and sunlit systems will be seriously degraded by the increase in wideband (large bandwidth) radio noise and consequent decrease in S/N. (Solar LF and MF bursts are usually more powerful than terrestrial transmissions and so penetrate the D-layer even when transmitted signals cannot.)

9.10 Shortwave Radio—HF Band

The 3-30 MHz frequency range is probably the most extensively used of all radio wave bands. If anything, frequency congestion is even worse on HF than for MF. Markedly greater bandwidth is available to HF operators, and HF easily provides the means of long distance communication. The basic technology for the employment of this band has long been available, and extensive research continues. The desirability of HF is tied up in its low cost and low power requirements combined with its truly worldwide range. The same physics which provides great range and good signal strength for low power also yields the major disadvantages of HF: highly variable propagation conditions (resulting in decreased reliability), high interference, and significant path variability.

A comparison of plasma frequencies and HF frequencies suggests that long-range sky wave propagation of HF will be by means of the F-region. The D-, E-, and F1-layers will refract (and occasionally reflect) the HF radio wave prior to its arrival in the F2-region. This means that more than purely geometrical considerations are necessary to determine the exact portion of the ionosphere illuminated by an HF signal. For example, the HF signal "spreads" as it moves outward from the antenna. At the F2-layer, a typical HF system will illuminate an area 10 to 20 km on a side. Ionospheric irregularities within the beam pattern will affect signal continuity by dispersing the signal. This reduces effective signal strength independent of other factors such as D-region absorption. For a given path, the workable frequency band lies between the maximum usable frequency (MUF) capable of being reflected by the ionosphere and the lowest usable frequency (LUF) still maintaining connectivity along the same path. In Figure 9.3, ray path 1 is associated with the LUF and ray path 3 corresponds to the MUF path. The LUF depends on the D-region losses and the changes to the signal path geometry as the ray passes through the E and F1 regions.

Systems equipped with directional antennas generally assume F2-layer reflection and great-circle propagation, but the ionosphere is not always cooperative. During daylight at solar maximum and occasionally during the summer daytime, F1 electron densities may exceed those of the F2-layer for a few hours. This is also observed during intense ionospheric storms. If elevation angle and azimuth are predicted on F2 geometry, communications may be lost during these periods. An extreme example of layer-induced variability is known as "skip." A term common to CBers (Citizen Band Radio; CB operates near 27 MHz) and ham operators, skip describes propagation at higher than normal frequencies over greater than normal distances. This is often a consequence of E-layer propagation. In particular, blanketing sporadic E may briefly double the range and MUF on a given path. For other users of the same path control point but with slightly different geometry, the occurrence of blanketing sporadic E may eliminate propagation for a time.

Since HF waves are reflected by the ionosphere, over-the-horizon (OTH) detection is possible. HF radars of this sort are often known as over-the-horizon-backscatter (OTHB) systems since they use backscatter from the target along the transmission path to the transmitter. (Note: the OTHB is not a backscatter radar; it forward scatters a signal to the target which is then reflected and returns to the radar by forward scatter.) Only a small portion (typically less than 1%) of the transmitted energy will be reflected back by the target. This means that a backscatter system requires considerable power to generate a detectable return. Furthermore, an OTHB system views not only its targets but also the Earth's surface and in order to differentiate between the two, radar electronics look for target motion.

9.11 Ionospheric Disturbances and Radio Propagation

As discussed in Chapter 8, sudden ionospheric disturbances (SIDs) are generally constrained to the sunlit hemisphere (though not always) as opposed to a fixed geographical locale. They are directly associated with energetic solar flares and are a consequence of the enhanced electromagnetic radiation of these flares. Although all solar wavelengths show enhanced emission during a flare, the largest are at x-ray and radio wavelengths. The primary impact of a SID is on the D-region, though other layers may be affected. Like the flare emission which produces it, the SID shows a sudden increase followed by a slow decline. D-region ionization is normally due to a balance between solar emissions and recombination, but during the onset of a flare this balance is briefly upset, and ionization climbs rapidly. The strongest SIDs usually result from large rapidly rising x-ray bursts. The radio burst associated with a flare may extend into the upper portion of the HF spectrum, and if the Sun is visible to an HF receiver antenna during such a burst, the receiver will experience a sudden increase in broadband radio noise. This will degrade the S/N ratio, perhaps below the point of usability. Since this effect works from the higher to the lower HF frequencies, it will affect frequencies near the maximum usable frequency first.

The shortwave fadeout (SWF) is the most classical of all HF SID effects. Since D-region absorption is inversely proportional to wave frequency, SWFs spread upward from the lower frequencies. A SWF may be sufficient to eliminate propagation on a particular path, or it may combine with a radio burst noise to limit the usable frequencies. The largest fades will occur for completely sunlit paths.

Not all ionospheric disturbances end within a few hours of their onset. Some persist for several days and cover large portions of the globe. These large-scale disturbances are loosely termed ionospheric storms (see Chapter 8). Some are related directly to the energetic particles produced by a large solar flare, while others occur in conjunction with large scale changes in the ionospheric electric fields in response to magnetospheric dynamics. As discussed in Chapter 8, solar induced storms include PCAs and geomagnetic storms.

A polar cap absorption event is a widespread, long-lived increase in D-region absorption confined to the polar ionosphere. Although the effects are similar in both poles, there is insufficient evidence to confirm or deny a one-to-one comparability, particularly in fine scale. High energy protons (greater than about 5 MeV) generated by a solar flare strengthen and lower the polar D-region resulting in the absorption of HF signals passing through the polar D-region. If proton fluxes possess sufficient energy density, they can shut down the HF propagation window and this phenomenon is known as a polar blackout.

PCA absorption typically begins as small patches and gradually expands to fill the polar cap. Absorption will usually be most intense in the sunlit polar cap (recall the ionosphere may be sunlit even when the surface is in darkness). The PCA will again become patchy as it dissipates.

Geomagnetic storms may profoundly alter ionospheric electron densities over large segments of the middle and high latitudes. The resulting impact on HF systems can show significant time variability. Ionospheric storms may occur in the absence of a discernable geomagnetic storm, but they are generally similar in their effect on HF systems regardless of the origin or association. The origin of a purely ionospheric storm is uncertain. As discussed in Chapter 8, a typical (geomagnetically induced) ionospheric storm will begin almost simultaneously with a geomagnetic disturbance and often will persist for some time beyond the end of the geomagnetic disturbance. The storm onset delay may range from 10 minutes in the auroral zone to 6–12 hours in the middle latitude evening sector following onset of the geomagnetic disturbance. The storm may persist for only a day at solar maximum to a week or more near solar minimum due to the restorative power of solar background emission at solar maximum. It is important to note that ionospheric effects of a geomagnetic disturbance are sometimes unrelated to the severity of the geomagnetic disturbance. Although a geomagnetically induced ionospheric storm can affect the entire ionosphere, only bottomside effects are significant to HF operations. Figure 9.8 shows how the MUF/LUF frequency band changes during ionospheric storms (see section 8.2). Storm characteristics are a function of local time, latitude, and season, and occasionally, significant differences are observed between the Eastern and Western Hemispheres as well.

9.12 Transionospheric Propagation

Ionospheric electron densities are usually insufficient to reflect frequencies above about 30 MHz which include the VHF, UHF, and SHF bands. These frequencies propagate by line-of-sight or possibly even sky wave if scattering by ionospheric irregularities is sufficiently intense. That is, the majority of the wave passes through the ionosphere but a small fraction of the wave "scatters" back downward and, in a sense, becomes something analogous to a sky wave, allowing communications over large geographic distances. The path irregularities introduced by scattering result in interference of mutually independent channels, and produce sufficient signal loss to limit the effective sky wave range to about 2000 km. System expense does not greatly exceed that of lower frequency bands, but VHF and higher bands provide considerably greater bandwidths.

Radar aurora is a radio wave phenomenon which affects both VHF sky wave and line-of-sight operations. Regions of enhanced auroral ionization produced by precipitating particles can reflect radar signals, but such radar aurora are not always found in one-to-one correspondence with visual aurora, nor would we expect it to be. Radar aurora, unlike optical aurora, is both ionization and aspect dependent. It has an average height of 110 km (close to the visual aurora) and varies from 75 to 135 km. Auroral radar returns are often

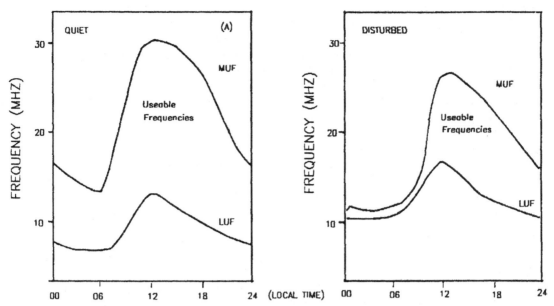

Figure 9.8 Diagram showing how the MUF/LUF varies from quiet ionospheric conditions (left) to ionospheric storm conditions (right) over the same signal path *(after Davies, 1965)*.

detected within ±20° of perpendicular to geomagnetic field lines. By reflecting a portion of the transmitted energy back to the radar, the radar aurora raises the noise level, or clutter, seen by the radar. The radar aurora also inserts Doppler shifts into its reflections, thereby adding additional problems to interpreting the radar returns. Auroral clutter may effectively jam some radars under certain conditions.

9.13 Satcom and Scintillation

Satellite communications (Satcom) systems use UHF and SHF (or above) bands to mitigate the ionospheric effects discussed earlier. In terms of frequency bands, communicators refer to UHF frequencies in and around wavelengths of 20 cm (1.5 GHz) as L-band frequencies; S-, C-, X-, and K-bands similarly correspond to wavelengths of at or near 10, 5, 3, and 1 cm respectively. However, these systems are not entirely immune to the ionosphere. The primary ionospheric effect on transionospheric systems is scintillation.

Scintillation is a rapid, usually random variation in signal amplitude and/or phase. Scintillation is thought to result from abrupt variations in electron density along the signal path which produce rapid signal path variations and defocusing as described in section 8.3. While variations over the entire path are important, the most significant variations occur near the F2-peak between 225 and 400 km. Scintillations cause both enhancements and fading in the radio signals; whenever the signal fades exceed the receiver's fade margin, the signal is temporarily lost. Of course, the impact of such signal interruptions will depend on the magnitude of the fades relative to the margin, the rate at which such fades occur, and the duration of the fades.

Satellite height is an important consideration in any scintillation analysis. A satellite orbiting at an altitude of 1,000 km sweeps past the F2-region irregularities at speeds up to 3 km/sec. In contrast, it is the normal ionospheric motions which carry the F-region irregularities across the signal path for geostationary satellites at speeds of about 100 m/sec. As a result, the scintillation rate observed from the signals from satellites in low Earth orbit (LEO) can be an order of magnitude greater than the rate observed for geostationary satellites.

Scintillation analysis is best done by geomagnetic latitude bands, with breaks at 20° and 60° from the equator. Recent work seems to suggest that scintillation is most clearly related to range spread; within a small horizontal distance, a particular electron density may exist at a variety of altitudes. Since the probe beam is of finite size when it hits the ionosphere, a portion of the wavefront will be returned from several different altitudes thereby affecting the down range distance the signal will go (thus the name range spread F). In contrast, frequency spread F may result from very strong electron density gradients capable of bending the incident ray, resulting in a poorly defined, or multiple values of the, critical frequency. Low-latitude scintillation is associated with F-region depletions as described in section 8.3, whereas high-latitude scintillation is associated with F-region enhancements.

Current research suggests that the effects of scintillation are most pronounced in the equatorial (±20°) latitude belt. Here, it seems closely associated with range spread F and attains maximum intensity between 2100 L and 0200 L. The onset is abrupt, with rapid, deep fading giving way to slower fading on a given frequency. The phenomenon may persist for 20 minutes to 2 hours at a given location.

Within the equatorial belt the occurrence and intensity of scintillation induced fade vary greatly with latitude. Near the geomagnetic equator, scintillation is generally weak, whereas near the equatorial anomaly, scintillation is the most intense found anywhere in the ionosphere. Figure 9.9 shows the monthly variation of scintillation measured by the satellite monitoring (1.5 GHz) station at Hunancayo, Peru, which is at the geomagnetic equator. Even though signal fading is most frequent near the spring equinox, the overall fade intensity is weak. Similar studies of scintillation at 137 MHz show a diffuse maximum extending from October to March and scintillation induced fade became practically nonexistent during the June solstice for quiet geomagnetic periods.

By comparison, Figure 9.10 shows the 1.5 MHz (L-band) fading due to scintillation measured at Ascension Island (about 17°S geomagnetic latitude); notice the fading is more frequent and more intense (20 kB as compared to 2 dB) than near the equator. In addition, both Hunancayo and Ascension show the same diurnal variation; that is, the scintillation is most intense from post-sunset up to midnight, after which the scintillation intensity weakens rapidly. Seasonally, all equatorial stations observe higher scintillation activity during the equinoxes, but stations in the Atlantic sector record minimum activity during the June solstice while Pacific stations appear to have a minimum during the December solstice. At this time there is no clear explanation of this seasonal anomaly, although some authors favor transequatorial wind flows and variations in the ionospheric conductivity as possible explanations for this seasonal behavior.

Geomagnetic activity reduces the occurrence of scintillation (and spread F) in the evening, but increases its presunrise intensity. Scintillation response to season, local time, and geomagnetic activity more closely mirrors range spread F changes than frequency spread F. Summer months generally result in minimum scintillation and a reversal of the impact of geomagnetic activity.

Scintillation patterns are poorly defined in the middle latitudes (20°–60° geomagnetic). Like spread F, scintillation is not common in this belt and shows no strong association with any single phenomena. A weak relationship seems to exist with geomagnetic activity, particularly at upper middle latitudes. Essentially anything which produces a sudden, sharp change in the ambient ionospheric conditions possesses the potential for triggering spread F scintillation. The lack of well-defined middle latitude patterns may not mean that such patterns do not exist because these latitudes have not been well monitored in the past.

High latitude (above 60°) scintillation shows two separate

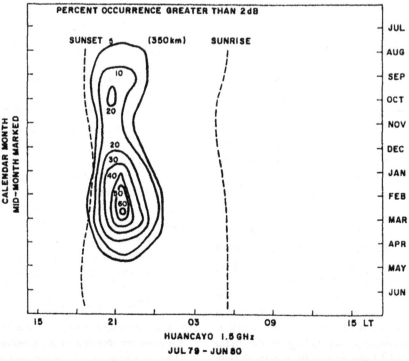

Figure 9.9 Schematic showing the percent occurrence of scintillation induced phase greater than 2 dB at 1.5 MHz as a function of month and local time. Measurements taken at Hunancayo, Peru (geomagnetic equator) during solar maximum for the period between July 1979 and June 1980 during geomagnetically quiet ($k_p < 3$) periods (*after Basu et al., 1988*).

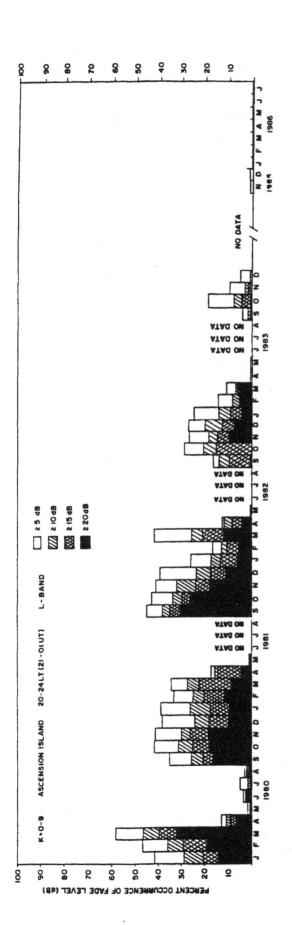

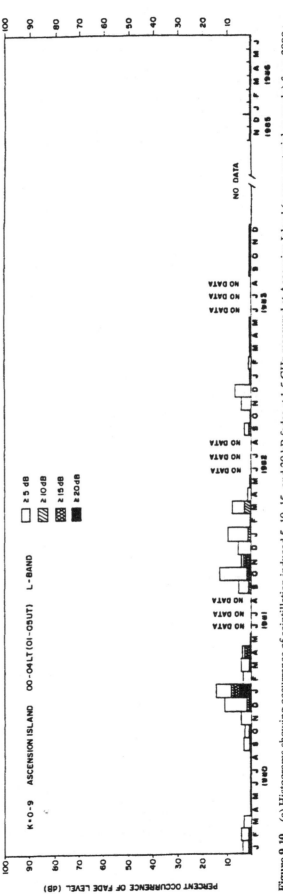

Figure 9.10   (a) Histograms showing occurrence of scintillation induced 5, 10, 15, and 20 kB fades at 1.5 GHz measured at Ascension Island (near equatorial anomaly) from 2000 to 2400 LT; measurements taken from 1980 (solar maximum) to 1986 (solar minimum) over all geomagnetic conditions. (b) Measurements for the same fade margins observed from 0000 to 0400 LT from 1980 to 1986 *(after Basu et al., 1988).*

centers of interest: the auroral oval and the polar cap. Particle precipitation patterns account for significant temporal electron content variation at a given point. The polar ionosphere is heavily dependent on cosmic ray and particle bombardment for its maintenance. Spread F is a ubiquitous consequence. The combination of these influences produces a structure somewhat aligned with the auroral oval. A scintillation "boundary" seems to exist about 5° equatorward of the oval itself. Scintillation increases poleward of the boundary to a maximum in the oval and a smaller secondary maximum over the magnetic pole. The subauroral trough is not coincident with the equatorward scintillation boundary. Moreover, hemispheric symmetry may not exist for this scintillation zone. The Southern Hemisphere boundary may occur several degrees higher in latitude than in the Northern Hemisphere. The boundary seems nearest the geomagnetic pole in summer and most distant in the winter hemisphere.

Figures 9.11 and 9.12 show the solar maximum scintillation behavior measured by the auroral monitoring station at Goose Bay, Labrador; Figure 9.11 shows the behavior for $K_p < 3.3$ (3+), whereas Figure 9.12 shows the measurements for the same time period but for conditions when $K_p > 3.3$. Based on the shape of the auroral oval and the fact the oval is oriented along the geomagnetic field (coordinates), Goose Bay is within the auroral oval only at nighttime between the hours of 2200 and 0400 magnetic local time. All measurements were made at 250 MHz from several polar orbiting satellites. As a result, the measurements are taken from satellite line-of-sight observations which cover the geomagnetic latitude range from 64° to 69°. Analysis of these observations show there was a higher probability of encountering scintillations at the higher latitude intersection points during geomagnetically quiet periods. This bias is associated with the location of Goose Bay relative to the auroral oval; during disturbed geomagnetic conditions the auroral oval can move south of Goose Bay toward the equator. Nonetheless, Figure 9.11 and 9.12 show that magnetic activity increases scintillation occurrences and magnitudes during both solar maximum and solar minimum.

Figure 9.13 shows the occurrence of 250 MHz scintillation fade measured by the polar cap monitoring station at Thule, Greenland, during solar cycle 21. The data shows the dramatic variation of both the scintillation magnitude and occurrence as a function of sunspot number. Since polar cap scintillation does not show a strong diurnal variation, the data in Figure 9.13 are 24 hour averages. Polar cap scintillation shows a general increase in intensity and occurrence with magnetic activity. Intense polar cap scintillation seems to be associated with the polar dayside cusp region. In addition, blobs of plasma convected away from both the dayside cusp and polar cap arcs are the most likely sources of the polar cap scintillation activity.

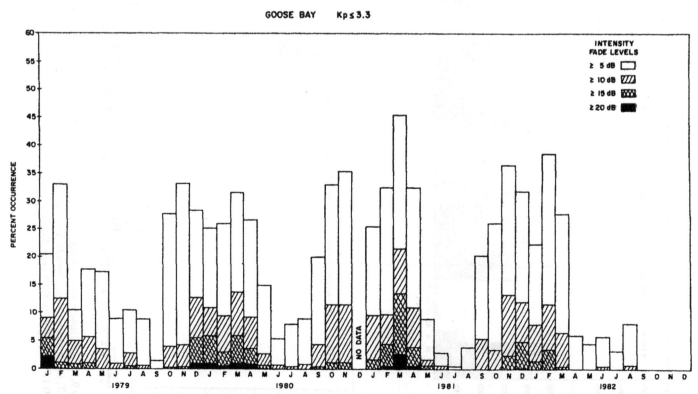

Figure 9.11 Histograms showing auroral fade induced scintillation measured at 250 MHz for geomagnetically quiet conditions ($K_p < 3.3$). Measurements made at Goose Bay during sunspot maximum (1979 to 1982); the apparent seasonal variation may be caused by the latitude changes of the transionospheric intersection point as described in the text *(after Basu et al., 1988)*.

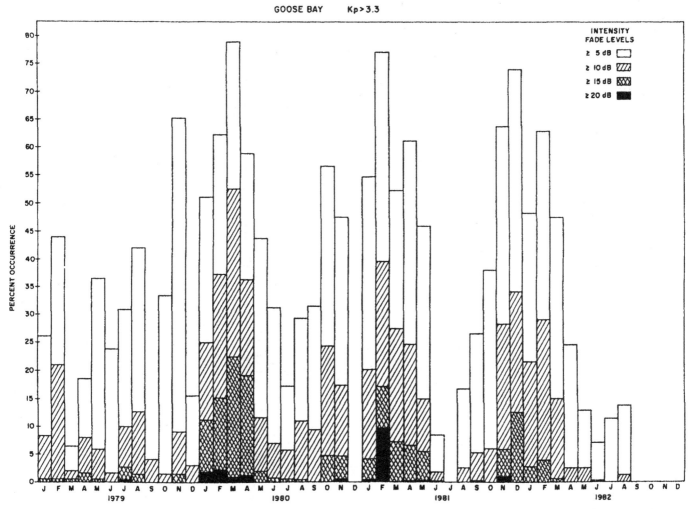

Figure 9.12 Goose Bay measurements for the same time period as in Figure 9.11 except for geomagnetically active periods (K$_p$ > 3.3); apparent seasonal variation may be an artifact of the measurement technique *(Basu et al., 1988)*.

Forecasting scintillation entails forecasting ionospheric anomalies associated with sharp ionospheric discontinuities. These include the sunset terminator, the auroral oval, and the geomagnetic anomalies, among others. Identifying areas and conditions conducive to sharp changes in ionization intensity or orientation is the most important step.

In summary, Figure 9.14 shows the global occurrence pattern for scintillation at 1.5 GHz (L-band) during both sunspot maximum and minimum. This figure shows three major scintillation regions: magnetic equator at post-sunset, nightside auroral oval, and the polar cap (including the dayside cusp) at all times of day. The most intense scintillation occurs in the equatorial anomaly region with more moderate levels of scintillation near the magnetic equator. At high latitudes, the scintillation activity is more moderate than in the equatorial region. Table 9.2 summarizes temporal and spatial characteristics of ionospheric scintillation.

9.14   Summary

Space operations make extensive use of the electromagnetic spectrum for communications, positioning, and detection. In some cases, the ionosphere facilitates these operations while in others it complicates operations. The EM spectrum is divided into bands loosely aligned with the type of interaction commonly expected between the radio wave and the ionosphere. Use of radio waves is predicated in part on their ionospheric interaction and in part on their available bandwidth and required equipment. ELF-VLF-LF systems operate by a waveguide mode to provide a truly global capability. Their utility is restricted by equipment expense and limited bandwidth. D-region height variations provide the major ionospheric impact at these frequencies. MF systems are generally short-range during daylight hours because of D-region intervention. At night, MF can cover considerable

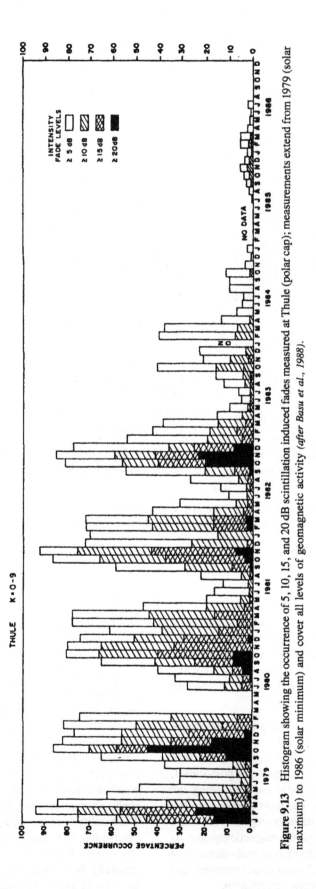

Figure 9.13 Histogram showing the occurrence of 5, 10, 15, and 20 dB scintillation induced fades measured at Thule (polar cap); measurements extend from 1979 (solar maximum) to 1986 (solar minimum) and cover all levels of geomagnetic activity (*after Basu et al., 1988*).

"WORST CASE" FADING DEPTHS AT L-BAND

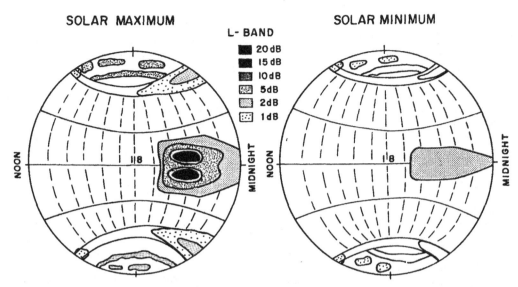

Figure 9.14 *Global occurrence pattern for scintillation at 1.5 GHz (L-band) during both sunspot maximum and minimum. Figure shows three major scintillation regions: magnetic equator at post-sunset, nightside auroral oval, and the polar cap (including the dayside cusp) at all times of day (after Basu et al., 1988).*

Table 9.2 Solar-terrestrial and temporal dependence of scintillation

	Latitudinal Range (geomagnetic)		
Parameter	±20°	20°–50°	60°–90°
Scintillation intensity	Greatest extremes	Slight to moderate	Moderate to extreme
Diurnal	All latitudes show maximum activity at night		
Seasonal	Maximum near equinox with slight longitudinal variations	Maximum in spring	Maximum in winter
Solar cycle	Increases with SSN	May be associated with rapid changes in magnetic activity level	Increases with SSN
Magnetic activity	Some longitudinal dependence	Increases slightly with magnetic activity	Increases with magnetic activity

distance by E-region refraction. HF is a truly long-range frequency band. Operating by F-region refraction, HF affords good bandwidth and inexpensive operation. Ionospheric variability is of primary importance to HF systems, since their operation is dependent upon it. VHF and higher frequency bands interact only slightly with the ionosphere—generally by scattering or scintillation. Conversely, they provide reliable circuits over primarily line-of-sight distances.

Understanding ionospheric radio wave propagation is really dependent on understanding ionospheric variability. Sharp ionospheric gradients, tilts, and transient inhomoge

neities are the origins of most operational concern. No system is immune to these irregularities, but an understanding of their effects will permit selecting the optimum frequency band for a particular location and operation.

9.15 References

Basu, S., Bureau, J., Rich, F. J., and Weber, E. J. 1985. Measuring Techniques—Ionospheric Radio Wave Propagation, in *Handbook of Geophysics and the Space Environment*, A. S. Jursa, ed., Air Force Geophysics Laboratory, National Tech. Info. Svs., Springfield, Va., p. 10-1.

Basu, S., MacKenzie, E. M., and Basu, Su. 1988. Ionospheric Constraints on VHF/UHF Communication Links During Solar Maximum and Minimum Periods, *Radio Science*, vol. 23, no. 3, p. 363.

Basu, S., MacKenzie, E. M., Basu, Su., Costa, E., Fougere, P. F., Carlson, H. C., and Whitney, H. E. 1987. 250 MHz/GHz Scintillation Parameters in the Equatorial, Polar, Auroral Environments, *IEEE J. on Selected Areas in Communications*, vol. SAC-5, no. 2, p. 102.

Bauer, S. J. 1973. *Physics of Planetary Ionospheres*. Springer-Verlag, New York.

Blanc, M. 1981. The Effects of Auroral Activity on the Midlatitude Ionosphere, in AGARD Conference Proceeding No. 295, *The Physical Basis of the Ionosphere in the Solar-Terrestrial System*. Pozzuoli, Italy 28–31 October 1980, Technical Editing and Reproduction, Ltd., London.

Bostrom, R. 1973. Electrodynamics of the Ionosphere, in *Cosmical Geophysics*, Egeland, Holter, and Omholt, eds. Universitetsforlaget, Oslo, Norway.

Chamberlain, J. W. 1978. *Theory of Planetary Atmospheres; an Introduction to Their Physics and Chemistry*. Academic Press, New York.

Davies, K. 1965. *Ionospheric Radio Propagation*, National Bureau of Standards Monograph 80. U.S. Government Printing Office, Washington, D.C.

Field, E. C., Heckscher, J. L., Kossey, R. A., and Lewis, E. A. 1985. Some Aspects of Long Wave Propagation, in *Handbook of Geophysics and the Space Environment*, A. S. Jursa, ed., Air Force Geophysics Laboratory, National Tech. Info. Svs., Springfield, Va., p. 10-20.

Hanson, W. B., and Carlson, H. C. 1977. The Ionosphere in *The Upper Atmosphere and Magnetosphere*. National Academy of Sciences, Washington, D.C.

Hargreaves, J. K. 1979. *The Upper Atmosphere and Solar-Terrestrial Relations*. Van Nostrand Reinhold Co., New York.

Holter, O., and Kildal A. 1973. Waves in Plasma, in *Cosmical Geophysics*, Egeland, Holter and Omholt, eds. Universitetsforlaget, Oslo, Norway.

Hunsucker, R. 1979. High-latitude E- and F-Region Ionospheric Predictions, in *Solar Terrestrial Predictions Proceedings*. U.S. Government Printing Office, Washington, D.C.

Klobuchar, J. A. 1979. Transionospheric Propagation Predictions, in *Solar Terrestrial Predictions Proceedings*, R. F. Donnelly, ed. U.S. Government Printing Office, Washington, D.C.

McNamara, L. F. 1985. High Frequency Radio Propagation, in *Handbook of Geophysics and the Space Environment*, A. S. Jursa, ed., Air Force Geophysics Laboratory, National Tech. Info. Svs., Springfield, Va., p. 10-45.

McNamara, L. F. 1992. *The Ionosphere: Communications, Surveillance, and Direction Finding*, Krieger Publishing Co., Malabar, Fla.

Maslin, N. M. 1987. *HF Communications: A Systems Approach*, Plenum Press, N.Y. and London.

Miller, A., and Thompson, J. C. 1975. *Elements of Meterology* 2nd Ed. Charles E. Merrill Publishing Co., Columbus, Ohio.

Mitra, A. P. 1979. Ionosphere-Reflected Propagation, in *Solar Terrestrial Predictions Proceedings*. U.S. Government Printing Office, Washington, D.C.

Moyer, V. 1976. *Physical Meterology*. Texas A & M University, College Station, Tex.

Nisbet, J. S. 1975. Models of the Ionosphere, in *Atmospheres of Earth and the Planets*, B. M. McCormac, ed. D. Reidel Pub. Co., Boston.

Prochaska, R. D. 1980. *Space Environmental Forecaster Course*. Air Force Global Weather Center, Neb.

Rawer, K. (1952). Calculation of Sky-wave Field Strength, *Wireless Engineer*, vol. 29, 287.

Rush, C. 1979. Report of the Mid- and Low-Latitude E and F Region Working Group, in *Solar Terrestrial Predictions Proceedings*. U.S. Government Printing Office, Washington, D.C.

Tascione, T. F., Flattery, T. W., Patterson, V. G., Secan, J. A., and Taylor, J. W. 1979. Ionospheric Modeling at Air Force Global Weather Central, in *Solar Terrestrial Predictions Proceedings*. U.S. Department of Commerce, U.S. Government Printing Office, Washington, D.C.

Thrane, E. V. 1973. Radio Wave Propagation, in *Cosmical Geophysics*, Egeland, Holter, and Omholt, eds. Universitetsforlaget, Oslo, Norway.

Thrane, E. V. 1979. D-Region Predictions, in *Solar Terrestrial Predictions Proceedings*. U.S. Government Printing Office, Washington, D.C.

Townsend, R. E. 1982. *Source Book of the Solar-Geophysical Environment*. AFGWC/WSE, Offutt AFB, Neb.

Vondrak, R. R. 1979. Magnetosphere-Ionosphere Interactions, in *Solar Terrestrial Predictions Proceedings*. U.S. Government Printing Office, Washington, D.C.

Wright, J. W. 1962. Dependence of the Ionospheric F-Region on the Solar Cycle. *Nature*, vol. 194, 461.

9.16 Problems

9-1.   Why is the ionospheric D-region a radio wave absorption layer? Is it an absorption layer for all radio frequencies? How about VLF?

9-2.   For radio waves, what is the index of refraction at the ionospheric reflection point? In a plasma, is the index of refraction greater than or less than one? If the index of refraction is less than one, do we violate the restriction on exceeding the speed of light? (Hint: differentiate between phase velocity and group velocity.)

9-3.   When using an ionosonde to determine the vertical electron density profile, does the ordinary wave or extraordinary wave propagate to higher "ionospheric altitudes"? Explain.

9-4.   How does an ionospheric sounding made with an incoherent backscatter radar differ from a traditional ionosonde? Can ionospheric soundings be made from satellites?

9-5.   Describe the propagation mode for the following types of electromagnetic waves: a) AM radio, b) FM radio, c) TV, d) high frequency radar (1–10 MHz), and e) satellite communications (300 MHz–1 GHz). List the advantages and disadvantages of each type of wave.

9-6.   How do ionospheric irregularities affect long-range HF propagation? Would the auroral oval affect HF radio waves following a transpolar propagation path?

9-7.   Distinguish between radio noise, fading, and absorption, listing the physical cause(s) for each phenomenon.

9-8.   Assume that a thick sporadic E layer and a thin sporadic E layer have the same peak electron density. Show the MUF for an oblique E-mode path reflecting from the thick sporadic E layer is less than the MUF for the same control point from the thin sporadic E layer.

9-9. Are shortwave fadeouts caused by SIDs? Explain. How about PCA events?

9-10. a. Starting with the differential form of Maxwell's equations, derive equation 9.2 for an electromagnetic wave traveling through an unmagnetized plasma. Assume the electromagnetic wave has the form $\vec{E} = \vec{E}_1 \exp\{i(\vec{k} \cdot \vec{x} - \omega t)\}$ and $\vec{B} = \vec{B}_1 \exp\{i(\vec{k} \cdot \vec{x} - \omega t)\}$ where E_1 and B_1 are the magnitudes of the electric and magnetic field components respectively and k is the propogation vector.

b. Assume the plasma now contains a uniform magnetic field $(\vec{B}_0)$ and the electromagnetic wave propagates perpendicular to $\vec{B}_0$ ($\vec{k}$ is perpendicular to $\vec{B}_0$). Show the index of refraction is given by equation 9.9. In addition, show that the ordinary wave propagates whenever $\vec{E}_1$ is parallel to $\vec{B}_0$, and that the extraordinary wave propagates when $\vec{E}_1$ is perpendicular to $\vec{B}_0$.

Chapter 10

Spacecraft Operations

10.0 Introduction

The growing importance of space operations has highlighted several geophysical problems. Spacecraft charging, drag variations, and radiation dangers to manned spacecraft are increasingly important features of the geophysical environment. For example, Explorer I discovered the trapped radiation belts and set off a furor of concern over the potential effects of radiation on spacecraft and astronauts. For many years, these concerns seemed generally unfounded, but the advent of more sensitive measuring devices and the switch to very low voltage integrated circuit technology has revived this concern. Over the past few years, it has become ever more apparent that the charged particle environment may affect spacecraft. This was clearly shown by the Jupiter Pioneer spacecraft. Its encounter with the energetic particles of the Jovian radiation belts nearly destroyed many onboard systems. This problem is not limited to interplanetary probes as was demonstrated by the Earth-orbiting ATS-6 satellite which recorded static surface potentials as high as 20,000 volts.

The long-term variations of extreme ultraviolet and soft x-ray emissions from the Sun change the amount of upper atmospheric heating which, in turn, affects the drag on low orbiting satellites. Geomagnetic storms are another source of heating. Although restricted to high latitude, the atmospheric response to these storms can alter the paths of polar orbiting satellites. The high energy particle precipitation accompanying a magnetic storm can also pose a significant radiation hazard to both manned and unmanned spacecraft. High intensity (hard) x-rays associated with some solar flares are another possible dangerous radiation source for high-altitude spacecraft. These and other problems relating to spacecraft operations are discussed in the following sections of this chapter.

10.1 Spacecraft Charging

Technically, spacecraft charging is a variation in the electrostatic potential of a spacecraft surface with respect to the surrounding plasma. Although the buildup of large static charges may confuse or blind certain sensors, the real danger lies in the resulting discharge because structural damage is a real possibility. Even weak discharges have been related to a variety of problems which include

1. Spurious electronic switching activity (such as turning off a recorder or activating a radio)
2. Breakdown of vehicle thermal coatings
3. Amplifier and solar cell degradation
4. Degradation of optical sensors

Although any vehicle operating above a few hundred kilometers may be susceptible to surface charging, the highest probability seems to be with geosynchronous vehicles. The high occurrence of charging on geosynchronous spacecraft may be an artifact of our reporting system, since a large number of high-altitude satellites are in geosynchronous orbit. However, theory does suggest that orbits above about $4 R_E$ should experience more problems because of their susceptibility to earthward streaming tail particles (see Figure 10.1).

Spacecraft charging is vehicle as well as orbit dependent. A spherical satellite with a homogenous, conducting surface would probably not experience significant charging-related problems because the vehicle's potential would be uniformly high. The utility of such a design is, of course, extremely limited. Nonetheless, vehicle design is an important consideration.

Two different mechanisms are thought to combine with vehicle design to generate spacecraft surface charging. Photoelectric effect and plasma bombardment are common terms for these culprits.

Illumination of the vehicle skin by photons knocks loose electrons. As these electrons are freed from the spacecraft (photoemission), the skin develops a relative positive charge. The electrons may form a negative plasma cloud or sheath near the vehicle skin. If the entire surface of the spacecraft were a homogeneous conductor, this charge buildup would generate a current flow to spread the charge evenly over the vehicle. Since most spacecraft exteriors have solar panels, probes, lenses, etc., there is a marked difference in conductivity across the surface. The result is differential charging of the sunlit surface with respect to the unlighted portions of the vehicle. Even in the best designed spacecraft, depressions or holes in the vehicle may be constantly shaded.

The success of plasma bombardment in charging a space-

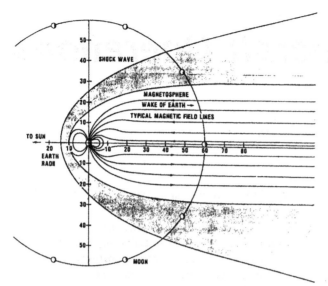

Figure 10.1 Cross section of the magnetosphere showing the approximate size in terms of number of Earth radii (1 Earth radius = 1 R_E = 6370 km). Geostationary orbits are at 6.6 Earth radii.

craft is structure dependent. A vehicle immersed in a hot (energetic) plasma is constantly colliding with charged particles. The extent and severity of surface charging depends, to a large extent, on spacecraft structure and design. Electrons with energies above a few keV are capable of penetrating 1 micron or more into a dielectric. Consequently, they "stick" to the spacecraft skin, causing a negative charge buildup. Holes or cavities in the front end of a vehicle (relative to its direction of flight) may actually scoop up energetic particles and accelerate this charging process.

Plasma bombardment can occur in several different ways. Vehicles at geosynchronous altitudes are susceptible to plasma injection events associated with geomagnetic disturbances and substorms. These events occur several times a day, even on quiet days, and may produce a ten-fold enhancement of ion density and a thousand-fold jump in electron density at geosynchronous orbit. As discussed in Chapter 5, particle injection occurs predominantly in the nighttime sector and results from an inward motion of the plasma sheet. The injected electrons drift eastward into the midnight-dawn sector, and the protons move westward into the evening sector. The result is a sudden immersion of high-altitude spacecraft in tail plasma. The greatest particle fluxes are to be expected slightly above or below the geomagnetic equatorial plane. As stated above, charging is most apparent near local midnight (spacecraft time) and is nearly invisible to vehicles operating on daytime meridians.

Geostationary vehicles are thought to be most susceptible to charging for two reasons. First, they are close to the magnetopause. In fact, on rare occasions they are outside the magnetosphere on the day side of the Earth. On the night side geostationary satellites can be, at times, immersed in the

night side plasma sheet, while vehicles below about 5 R_E probably are not. Second, the ambient plasma density at 6.6 R_E is low. This means that, unlike low orbit vehicles, the ambient atmosphere is incapable of "bleeding off" or neutralizing small charges before a discharge can occur. For low orbiting vehicles, significant charging requires unique conditions but is not, unfortunately, impossible.

It would seem that the photoelectric effect and plasma bombardment might at least partially offset each other. To some extent, they seem to do so. The difficulty is again in vehicle design. What happens on the surface of a spacecraft is a sort of balancing act. All currents (positive and negative) to and from the surface must balance. In order to obtain this balance, the surface potential (voltage) must vary. Some parts of the craft will, therefore, generate higher potentials than others.

Spacecraft charging sets the stage for subsequent discharges. Since discharges are capable of producing greater lasting damage, it is important to identify those conditions conducive to discharging. Any time charging is occurring, conditions are favorable for discharge. Experience indicates sudden changes in the electrical environment of the spacecraft may trigger static discharges. The simplest triggers include orbital maneuvers, the onset of downlink telemetry, or other electronic activity onboard the spacecraft. Eclipse or movement out of eclipse of a geosynchronous vehicle (happens near equinox) or the movement into/out of sunlight for a lower orbit vehicle may also trigger a discharge. Encountering an intense current or the boundary of the magnetosphere can also be triggering possibilities for high-altitude spacecraft. Anything which can cause a sudden change in the distribution of the vehicle potential is a possible triggering mechanism.

Initial studies have suggested some statistics for analyzing (or forecasting) discharge times for geostationary satellites. During low magnetic activity, discharges seem more common between 0400 and 0600 L (spacecraft time). This may be due to quiet time injection events and the preferred direction of drift for injected electrons (eastward). Noon local is a favored discharge time during high magnetic activity, perhaps due to the increased likelihood of encountering the magnetospheric boundary (or associated current systems). Early evening (1900 L) seems to yield the minimum probability of discharges. The equinoxes bring eclipses for geosynchronous vehicles, and simultaneous increases in discharge probability.

While we can, with some success, forecast the environmental "conduciveness" for spacecraft charging, we are not yet at that point for discharges. Charging is heavily dependent on the environment for all vehicles. Discharging is perhaps more dependent on vehicle design and depends on the environment in unknown percentages and ways. Nonetheless, a rapid postanalysis of a spacecraft anomaly may be very useful to an operator. If a problem can be traced to spacecraft charging, it may preclude the expense (in terms of money and data lost) of vehicle shutdown and subsequent testing. A better understanding of vehicle design requirements and the environment in which the vehicle must operate may help engineers design more discharge-resistant vehicles. A low voltage solid-state circuitry vehicle has been flown (SCATHA, Spacecraft Charging at the High Altitudes) solely to test various methods to minimize changing effects.

10.2 Deep Dielectric Charging

Electrons with energies between 2 and 10 MeV have enough energy to burrow deep into satellite surfaces. This excess charge spreads out evenly on conducting surfaces, but the charge accumulates on dielectric surfaces resulting in uneven electric potentials between different portions of the satellite. Eventually, potential differences can reach the breakdown threshold and a static discharge will then occur. Unlike surface charging, deep dielectric charging can form strong potential differences on the inside surfaces of satellites, and thus the resulting discharges can arc directly into the satellite's internal electrical circuitry.

An example of a deep dielectric anomaly is shown in Figure 10.2. The inability of a certain satellite's star tracker to maintain a track lock shows a strong correlation with the arrival of 3 MeV electrons. Notice also the correlation is not

perfect because on three instances the electron flux seemed to cross the sensitivity threshold without disrupting the star tracker. There are two possible reasons for such behavior. The first possibility is the satellite design; satellite electronics are often clustered and sensors are rarely evenly distributed. Thus, there are times when sensitive electronic components are shielded from direct particle interactions by other portions of the satellite. Second, the high-energy electron measurements were made on a satellite which was close to the satellite experiencing star tracker problems, but nevertheless the two satellites were separated by an appreciable distance. Therefore the environment sampled by that satellite may not be representative of the environment surrounding the other satellite.

10.3 Single Event Upsets

Single event upsets (SEUs) are bit flips in digital microelectronic circuits. SEUs can cause:

a. Damage to stored data
b. Damage to software
c. The central processing unit (CPU) to halt
d. The CPU to write over critical data tables
e. Various unplanned events due to faulty commands

SEUs in spaceborne electronics are caused by the direct ionization of silicon material by a high energy ion passing

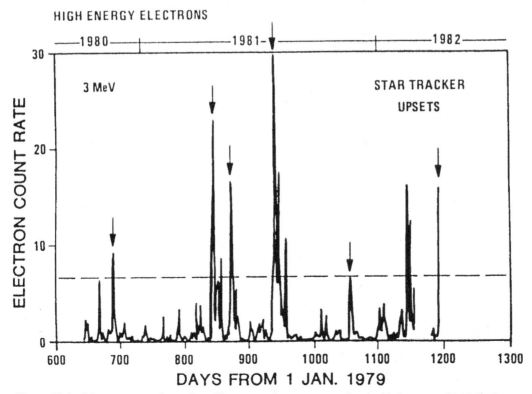

Figure 10.2 Measurements of actual satellite star tracker upsets correlated with the onset of 3 MeV electron fluxes; dashed line represents apparent threshold for star tracker upset to occur *(after Baker et al., 1986)*.

through it. The near Earth particle environment includes galactic cosmic radiation (GCR), energetic particles from the Sun, and trapped protons. The normal factor in SEU production is the heavy ion cosmic ray, although large solar flares can produce a substantial increase in SEUs. Fortunately, such large flares occur only once every few years. Figure 10.3 illustrates the variation in heavy ion GCR over the 11 year solar cycle. For satellites in near Earth orbits (less than four Earth radii), an additional factor is the radiation belts, which trap and hold high energy charged particles. A satellite passing through these belts is subjected to a high flux of electrons and protons, and trapped protons are capable of causing SEUs in certain types of chips.

Figure 10.4 depicts the cosmic ray flux as a function of energy deposition in silicon. The deposition curve is plotted for several thicknesses of shielding. Notice that shielding of 20 g/cm^2 reduces the flux by less than a factor of 10 compared to the standard satellite shielding of 2 g/cm^2 provided by a typical skin plus electronics box. Clearly, from Figure 10.4, a factor of 10 decrease in occurrence is not a great victory since an upset every 15 days is not much better than one every 1.5 days in the life of a 10-year system. Also, the weight and volume of 20 g/cm^2 shielding, 3.0 inches of aluminum, is not a very reasonable solution.

The upset rates shown in Figure 10.5 are the results of a calculation which depends on measured data for parts and a standard environment. The environment is specified as the 10% environment meaning that the actual environment is worse than this only 10% of the time. The regions labeled CMOS-BULK and CMOS-EPI reflect the test data. This chart shows the strong dependence of upset rate on both feature size, stated as the collection volume, and the critical

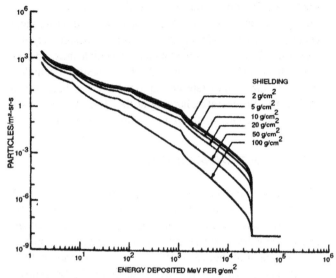

Figure 10.4 Cosmic ray flux as a function of shielding *(after Adams, 1981).*

charge. Critical charge is the minimum accumulated charge required to produce an SEU. At geosynchronous altitudes all three particle sources contribute to the SEU rate plotted in Figure 10.5. The SEU effectiveness of heavy ions versus protons is about 10^5 to 1.

Four methods or techniques are in use or planned for use on satellite electronics to mitigate the impact of SEUs. Satellite designers must evaluate these methods in order to trade off the penalties and develop a protection scheme while still meeting the system performance requirements.

Clearly, avoiding the problem through parts selection looks attractive. In fact, many older technologies such as plated wire, or core or solid state memory with large feature size element (>10 microns) can be used to achieve a small, or zero, upset rate. The problem here is usually operating speed

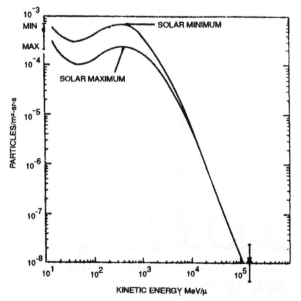

Figure 10.3 The IRON cosmic ray environment showing the variations due to solar activity.

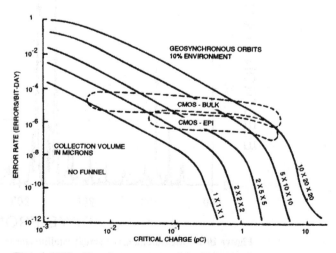

Figure 10.5 Cosmic ray upset rates *(after Cunningham, 1984).*

and power consumption. The parts may not meet design or system performance goals. Various investigators have designed and built upset-resistant parts while maintaining circuit performance.

Triple memory redundancy can beat the speed problem as well as provide a high degree of fault tolerance from other causes, such as parts failure. Negative factors include the additional weight and power required for the additional memory. Protection must also be provided for the common "voting" electronics.

Error detection and correction (EDAC) offers the advantage of a single memory, but with additional bits stored to check for unexpected bit flips (i.e., parity errors). Also, the processor and software must perform EDAC on some sort of regular basis to ensure a correct memory. As upset rates increase, the point could quickly be reached where most of the time is spent doing EDAC rather than any productive work. One current design has a predicted upset rate of slightly less than one per day. In this case, since EDAC is planned for several times per minute, the chances favor quickly detecting and correcting the upset.

The final method of dealing with SEUs is to regularly reset the onboard computers. A variation of this is to use a regular sequence when commanding the satellite; any change in the sequence would be easy to detect and correct. Computer systems which can be reprogrammed from the ground might allow other alternatives. This method is usually unsatisfactory, unless an occasional reset is acceptable within the system performance requirements.

10.4   Spacecraft Drag

Spacecraft operating below a few thousand kilometers encounter a significant number of atmospheric particles during each orbit of the Earth. Theoretical atmospheric models permit us to calculate the expected drag and its effects on vehicle orbit. Changes in atmospheric density at the vehicle altitude can rapidly and significantly alter the vehicle orbit. If these are unexpected changes, the vehicle may be temporarily "lost" by ground tracking stations using narrow-beam tracking radars. This happened several times during the Space Shuttle flight in Spring 1981. In order to eliminate the unexpected nature of such changes, we must identify their origin/cause and attempt to forecast it.

Any mechanism capable of heating the Earth's atmosphere will produce density changes at altitudes above the level heated. Such heating may result from geomagnetic storms and changes in solar EUV (extreme ultraviolet) emission.

During a geomagnetic storm, large numbers of particles are dumped into the high-latitude atmosphere. These particles generate considerable heating near 100 km and alter densities to 1000 km altitude or higher. Heating effects (i.e., drag) are first observed a few hours after the disturbance

begins and may persist for 12–24 hours following a large disturbance. The strong field-aligned currents and the enhanced electrojets contribute to atmospheric heating. Most of this heating will be near the auroral zone, and consequently, polar orbiting spacecraft will experience the greatest effects from geomagnetic storm heating. As mentioned in Chapters 5 and 8, auroral substorms produce a well-organized ionospheric, two-cell, horizontal circulation pattern across the polar cap. Ion-neutral collisions spin up a similar circulation pattern in the neutral atmosphere above a few hundred kilometers. The resulting "neutral winds" can attain speeds in excess of 1 to 2 km/s which, in turn, can make a significant impact on the *total* drag experienced by low-altitude, polar orbiters moving at speeds of 6 to 7 km/s.

The greatest long-term impact on atmospheric density probably results from heating by solar EUV fluctuations. Since EUV measurements are not available in real time, daily measurements of 10.7 cm radio emissions are taken to be representative of solar EUV flux. Generally, solar emissions vary daily, with a 27 period, and over the solar cycle. As mentioned earlier, EUV (and 10.7 cm) daily and monthly variations seem to be related to changes in plage area on the solar disk. Generally, day-to-day atmospheric density variations are masked by long-term trends and occasionally, density variations are reinforced by geomagnetic activity.

All variations in density result in variations in atmospheric drag on low-altitude (below 1000 km) satellites. Any orbital body in an atmosphere experiences a drag which causes changes in its orbit. Since drag is greatest at perigee, the decrease in speed is most significant there. In fact, to first order approximation, the decrease of velocity at apogee for highly elliptical orbits may be ignored. The decrease in perigee velocity results in a decrease in apogee height, and the orbit will slowly become less elliptical and more circular. When the perigee and apogee become sufficiently close in height, drag becomes significant at all points along the orbit, and the orbit quickly decays.

Density need not steadily increase to cause problems. A single increase will alter not only the next orbit, but also all subsequent orbits. Likewise, an unexpected *decrease* in density can cause problems. Thus, forecast centers use the geomagnetic indices and 10.7 cm radio flux observations (and forecasts) to change their atmospheric density model. This, in turn, is used to correct the expected locations of satellites for some future time and to plan orbital maneuvers.

Specific comparisons for a satellite at 185 km can be made in terms of resulting in-track displacements (how far ahead or behind it is compared to its calculated position). Over 12 hours, a $K_p = 3$ produces a 5 nmi (nmi = nautical mile) displacement, while a $K_p = 8$ produces a 50 nmi displacement. The effects of geomagnetic activity will be most severe on polar orbiting vehicles. Moreover, density variations resulting from geomagnetic activity may persist for 8–24 hours after the end of the disturbance as measured by ground-based magnetometers. Since the level of geomagnetic activity varies rapidly, atmospheric density models must be continually updated during disturbed periods.

Over 3 days, a 10.7 cm flux level of 70 solar flux units (solar minimum) results in a 600 nmi in-track displacement at 185 km (compared to orbit with no drag effects). A 220 (solar maximum) flux level produces an 1800 nmi displacement in 3 days for a vehicle at 185 km. The severity of atmospheric heating was demonstrated by the early demise of Skylab which succumbed to the increase drag during solar maximum.

10.5 Space Shuttle Glow

In the early 1980s, Space Shuttle astronauts observed an orange glow on the portion of Shuttle's exterior surface facing into the direction of motion which is commonly called the ram direction. Observations indicate the glow fades with increasing distance from the Shuttle with an exponential decay length of about 20 cm. Although the glow is not completely understood, researchers agree the glow is the result of the interaction of the Shuttle's exterior with the ambient atmosphere.

Spectral measurements show the glow emissions are continuous in and around 600 nanometers with a peak near 680 nanometers. This spectrum resembles the emissions expected from NO_2, although measurements at longer wavelengths show evidence of OH emissions. At Shuttle altitudes there doesn't appear to be enough natural NO_2, or OH, to account for the glow intensity. Therefore, current theories on Shuttle glow favor the idea that the necessary concentrations of NO_2, and OH, are manufactured as the Shuttle surface collides with the ambient atmosphere.

One widely accepted theory postulates that NO_2 is formed in a multistep process, starting with thermospheric atomic oxygen chemically reacting with the nitrogen in the material making up the Shuttle's surface. The resulting NO_2 halo surrounds the Shuttle with the newly manufactured NO_2 molecules in excited vibrational states.

Other theories suggest the glow originates from chemical interaction of thermospheric atomic oxygen with molecules being outgassed from the Shuttle. As might be expected, this theory predicts the glow intensity is a function of the amount of outgassed molecules. Since the Shuttle bay is the most significant source of outgassing, the glow should vary as the Shuttle bay moves in and out of the ram direction during routine Shuttle maneuvers. Measurements seem to show the variation in glow intensity is inconsistent with the predictions based on outgassing as the only source of the glow. Therefore, the source of the glow is probably a combination of outgassing and surface interactions.

A glow also appears when the Shuttle fires its maneuvering thrusters. This thruster induced glow is more intense and extensive than the usual surface glow. The thruster glow is produced when the thruster combustion products chemically react with the ambient atomic oxygen. However, observations show the glow sometimes persists longer than the thruster firings. The answer to this puzzle seems to be due, in part, to the attachment of some of the thruster combustion products to the Shuttle's exterior surfaces; these products include NO, N_2, and CO_2, as well as more complex hydrazine molecules. In turn, the resulting surface coating chemically interacts with the colliding thermospheric oxygen atoms to form a surface glow.

10.6 Space Radiation/Manned Space Flight

Although the study of space radiation began early in this century, most of our knowledge was acquired during the last three decades. The advent of manned space flight provided an immense number of observations as a basis for analytical treatment of space radiation hazards. While research efforts have provided a basic understanding of these hazards, a great deal remains unknown. We are constantly bombarded by many types of ionizing radiation. Fortunately, our atmosphere blocks much of the harmful component, but in space, this protective shield is not available.

The main hazard to life in space is found in the ionizing radiation resulting from exposure to high energy particles. If a particle possesses sufficient kinetic energy it can pass through protective equipment and impact a crew member's body. Particles of very high energy may pass through an individual with *no* serious effects. It is the particles which are stopped by human tissue that pose the most danger. As these particles rapidly decelerate, their energy is converted into a pulse of electromagnetic (EM) radiation. This radiation can ionize atoms within the crewmember's body. In addition, the impact of the energetic particle can excite neighboring atoms. As these atoms fall back to lower energy states, they too can produce ionizing radiation.

Natural radiation in near Earth space (up to geosynchronous orbit) has three primary components: galactic cosmic radiation (GCR), radiation produced by trapped particles and radiation from solar flare particles. All three components are influenced by solar activity and the Earth's magnetic field. Their relative contributions to radiation hazards are most easily understood when considered separately.

GCR is high energy (greater than 0.1 GeV) protons, electrons, or other heavy, energetic particles. Emitted by distant stars and even more distant galaxies, they diffuse through space and arrive at Earth from all directions. Spatial variations in GCR flux (and therefore GCR related radiation) are produced by variations in source location, the Earth's magnetic field, atmospheric shielding, and with increasing altitude. Particle flux is also larger over the polar regions where "open" geomagnetic field lines allow easier access.

The most important temporal variation in flux is associated with the 11 year solar cycle (see Figure 10.3). During solar maximum, when the interplanetary magnetic field strength is greatest, cosmic ray particles are scattered away from the Earth. This produces a GCR flux minimum (Forbush decrease). Conversely, GCR flux is largest during solar minimum. The 11 year solar cycle produces a factor of two variations in the cosmic ray dose at a geosynchronous orbit. Low-altitude, low-inclination orbits would experience almost no dose variations due to the strong shielding produced by the combined effects of the atmosphere and geomagnetic field.

Due to its extremely high energy, a GCR is very penetrating, and spacecraft shielding is not very effective in reducing the radiation dose. Fortunately, GCR flux is comparatively low, so it doesn't pose a serious threat to humans (several particles have probably passed through your body since you started reading this section). In all orbits, approximately 5–10% of the total effective radiation dose is due to GCR. This small amount is sometimes referred to as background radiation.

Energetic protons trapped in the inner radiation belt are the major source of radiation for Earth orbiting spacecraft above 500 km, particularly in the South Atlantic anomaly region. The amount of radiation varies with latitude and longitude (the inner belt extends to about 45° latitude). The inner belt proton population is also susceptible to solar-induced variations. Population density varies out of phase with the 11 year solar cycle, so that the inner belt is most inflated during solar minimum. This variation in particle population produces a factor of two variation in radiation dose rate during the solar cycle for low orbiting spacecraft.

The outer radiation belt is asymmetric, with the nightside being elongated and the dayside flattened (see Figures 4.7 and 4.8). Generally, particle energy and outer boundary location vary with the 11 year cycle. During solar maximum, the outer boundary of the electron belt is closer to the Earth and contains higher energy particles. At solar minimum, the outer boundary moves outward and contains fewer energetic electrons. Outer belt electron densities undergo order of magnitude changes over time scales of weeks. These short-term variations can produce significant radiation dose variations and are related to the level of geophysical activity. During, or shortly after, very active periods, the outer belt is inflated with high energy electrons which increase the radiation hazard substantially. Diurnal variations in radiation dose inside a spacecraft (in high-altitude circular orbits) can occur when the trajectory crosses the asymmetric outer electron belt.

These trapped radiation belts present a serious threat to the space traveler. The locations of the most hazardous regions are sufficiently well known so that flight trajectories may be planned to limit time spent in these areas. For mission confined to these dangerous regions, close monitoring of the radiation dose and acceptable dose accumulation must be specified (see Figure 10.6).

Solar protons, also referred to as solar cosmic rays (SCRs), represent the third and most variable component of natural space radiation. Solar cosmic rays are composed of protons, electrons, or other heavy nuclei accelerated to energies between 10^7 and 10^9 eV during very large solar flares. These particles can be responsible for a thousandfold increase in the radiation dose over short periods of time. Similar to the other energetic particles, SCRs produce ionizing radiation when they interact with atoms (shielding or body tissue).

Solar particle events show a correlation with the 11 year solar cycle. The largest events normally occur in the months following sunspot maximum. Individual events vary considerably in particle constituents, energy spectra, and particle flux. Usually, a few very large flares dominate the total particle fluence for the entire solar cycle. Due to the "individuality" among events, the radiation dose accumulated due to solar protons may vary from negligible to well above lethal. Figure 10.7 shows the frequency of solar proton flares during a solar cycle.

For a fixed altitude, spacecraft can experience different levels of radiation depending on orbit trajectory. Equatorial orbiting spacecraft will experience lower proton fluence (and therefore a lower radiation dose) than a polar orbiting satellite at similar altitudes. In general, solar proton radiation is a significant hazard for orbits passing above 50° latitude at altitudes above a few Earth radii (1 Earth radius = 6378 km = 3960 miles). Solar cosmic rays emitted during a large solar flare present the greatest uncertainty and the greatest threat to manned spacecraft in regions beyond the protection of the Earth's atmosphere.

The hazards posed by trapped particles and high energy flare particles are difficult to assess in a quantitative manner. The deleterious effects of radiation depend on total dose, dose rate, particle identity, and energy, or a combination of all these variables.

10.7 Radiation Health Hazards

When high energy particles encounter atoms or molecules within the human body, an atomic interaction (ionization) may occur. A direct interaction occurs when the particle is suddenly stopped by collisions resulting in a release of energy which may remove electrons from nearby atoms or molecules, and ions result. Indirect encounters occur when the high energy particle, usually an electron, is deflected by another charged particle. The deflection causes a release of energy (radiation) which also may produce ionization. The close encounter process is commonly referred to as Bremsstrahlung. In either interaction, the effects of the ionizing radiation are proportional to the amount of energy absorbed by the surrounding material. Ionizing radiation can break chemical bonds in biological molecules. These, in turn, cause molecular changes which can result in biological injury.

To quantify this absorbed radiation, a unit of measurement called a rad was defined. A rad is the amount of ionizing radiation corresponding to 0.01 joule absorbed by one kilogram of material. Note that a dose of 10 rads from high energy protons is the same as 10 rads from x-rays. The rad represents an amount of absorbed radiation energy and not the source or type of radiation. Regardless of the source, a sufficiently large dose of radiation will kill any living organism. About 100 rads of radiation will have the immediate effect of causing radiation sickness in humans. Radiation scientists often use a unit called a gray (1 Gy = 100 rads) when discussing the effects of radiation on humans. For example, an x-ray dose of 1 to 2 Gy has a high probability of killing a cell by producing dozens of lesions in a cell's DNA molecules.

Radiation physics indicates that 1 rad received from x-rays produces far less bodily damage than 1 rad received from high energy protons even though both deposit equal amounts of energy. Another unit was developed to express the effects of radiation on biological tissue. To define this unit, each type of ionizing radiation was given a relative biological effectiveness (RBE) compared to a beam of 200 keV x-rays. Table 10.1 lists the RBE estimates for several types of radiation. From the table, we see that protons can be twice as damaging as 200 keV x-rays, and therefore, a 1 rad proton dose will be twice as damaging as a 1 rad dose from x-rays. A rem relates biological damage to type of radiation (rem = dose [rad] × RBE). For example: a 1 rad dose of 200 keV x-rays gives a biological equivalent dose of 1 rem, but a 1 rad dose from protons gives a biological equivalent dose of 2 rem. The larger rem value for protons accounts for the increased biological damage. Accumulated data suggests that electrons, protons, neutrons, and alpha particles are the most damaging (largest RBE value) due to their ability to penetrate deeply into human tissue and release or produce a large number of ions. Analogous to the gray radiation unit, scientists sometimes use the unit called a Sievert (1 Sv = 100 rem) to describe human biological damage from radiation.

On the average, we experience about 40 millirems (mr) each year from radioactive elements in soil, rock, and wood around us. This figure varies from place to place. On the East Coast, it is around 20 mr; near the Rocky Mountains, the value is closer to 90 mr. Cosmic rays passing through your body provide an additional 40 mr annual dose (160 mr if you live high up in the Rocky Mountains). Inescapable sources in food and water provide an additional 20–50 mr which brings the total yearly dosage for an earthbound person to about 170 mr (add 4 mr for each New York to Paris airline flight).

In contrast to this, the space traveler will receive considerably more radiation. The dose received varies with mission duration, orbital profile, and shielding. Table 10.2 highlights some of the radiation doses experienced by Gemini astronauts during the mid-1975 period (solar minimum).

Shielding stops or alters the trajectory of high energy particles before they encounter the more sensitive human tissue. In general, the denser a material the more effective it is as shielding. Unfortunately, with booster limitations on weight launched, denser shielding means reducing payload elsewhere. Aluminum is used extensively, since it combines both high density and lightness. A typical space suit only contains a small amount of aluminum which effectively stops up to 10 MeV protons. Higher energy particles pass through the suit and can deposit most of their energy in the individual. Spacecraft exteriors typically have several grams per cm^2 of aluminum shielding and can stop even higher energy particles. Table 10.3 reflects the doses expected for various orbits and shielding thicknesses.

Low orbits, especially those confined to the equatorial plane, are substantially less hazardous than polar orbits. The former make use of the Earth's natural shielding, while the latter expose an individual to ambient energetic particles in a region where natural shielding is of limited value. Shielding is particularly important at geosynchronous orbits.

Models employed in predicting doses are usually based on previously measured particle fluxes for a given environment. Unfortunately, methods used to convert particle flux into a biologically equivalent dose are often limited by assumptions about particle types and energy range, flux density, and shielding effectiveness. One model assumes the most dam-

Table 10.1 Relative biological effectiveness of various radiation sources *(Bueche, 1981).*

Radiation	RBE
5 MeV gamma rays	0.5
1 MeV gamma rays	0.7
200 keV gamma rays	1.0
Electrons	1.0
Protons	2.0
Neutrons	2–10
Alpha particles	10–20

Table 10.2 Gemini orbital parameters and average radiation doses in millirads *(after Atwell, 1980).*

Mission Launch Data		Apogee (NM)	Perigee (NM)	Revs	Inclination (deg)	Avg Dose Millirad
I	Aug 21, 1965	189	87	120	32.5	176
VIII	Mar 16, 1966	161	86	7	29	10
IX	Jun 3, 1966	168	86	45	29	19
X	Jul 18, 1966	412	161	8	29	726

Table 10.3 Predicted radiation dose (rem) for a 90 day mission assuming an RBE of 1 *(after Bostrom et al., 1987)*.

Orbital Inclination/Altitude	Shielding Thickness	Predicted Dose
28.5°/450 km	1.0 gm/cm^2	
Trapped radiation		7.3 rem
GCRs		0.4 rem
Solar Flare Cosmic Rays from an Anomalously Large Flare (SCRs/ALF)		0.0 (shielded by geomagnetic field)
57.0°/450 km	1.0 gm/cm^2	
Trapped radiation		4.7 rem
GCRs		0.7 rem
SCRs/ALF		40 rem (skin)
		4 rem (blood-forming organs (BFO))
90°/450 gm	1.0 gm/cm^2	
Trapped radiation		4.2 rem
GCRs		0.9 rem
SCRs/ALF		420 rem (skin)
		29 rem (BFO)
Geostationary Orbit	2.0 gm/cm^2	
Trapped radiation (electrons)		4.3 rem
GCRs		1.8 rem
SCRs/ALF		1100 rem (skin)
		105 rem (BFO)

aging particles from a solar flare are protons having an average energy near 50 MeV. This method makes several restrictive assumptions to yield a very rough estimate. The limited accuracy of these models makes them useful for mission planning but not practical for monitoring daily space operations. For safety and greater accuracy, daily operations must rely on in situ measurements of accumulated dose and dose rates. Figure 10.8 illustrates fluctuation in daily dose rates for a high-altitude orbit with a 12 hour period.

To further complicate the difficulty in assessing biological damage for a given dose, studies show that individuals have varying degrees of tolerance based on sex and stamina. To circumvent this problem, most estimates of radiation effects are based on a population sample with the number of affected individuals expressed as a percentage. Table 10.4 relates accumulated dose to probable prompt effects on a sample population. The table does not include long-term effects such as cancer.

10.8 Radiation Hazards to Satellite Electronics

High energy particles degrade detector performance by the accumulation of material microstructural damage. Different devices have varying degrees of total dose vulnerability which can range from very soft (700 rads) to very hard (10^6 rads). Figure 10.6 shows the accumulated radiation dosage for a five year mission due to trapped radiation particles for a variety of orbits. Also plotted are the typical shielding thicknesses available from the satellite skin and electronic boxes.

Table 10.4 Probable radiation dose prompt effects for a sample population *(Cladis et al., 1977)*.

Dose (rem)	Probable Effect
0–50	No obvious effect except, possibly, minor blood changes.
50–100	Radiation sickness in 5–10% of exposed personnel. No serious disability.
100–150	Radiation sickness in about 25% of exposed personnel.
150–200	Radiation sickness in about 50% of exposed personnel. No deaths anticipated.
200–350	Radiation sickness in nearly all personnel. About 20% deaths.
350–550	Radiation sickness. About 50% deaths.
1000	Probably no survivors.

Besides trapped particle radiation, geostationary and polar orbiting satellites can experience a substantial radiation dose from solar flares. Figure 10.7 shows the frequency of solar proton flares over a typical solar cycle. Protons with energy above 30 MeV are a hazard to satellite electronics. As mentioned above, Figure 10.8 illustrates how the solar proton dose can vary from day to day.

10.9 Mission Planning/Safety

Recognizing that no region of space is free from radiation, mission planners are forced to accept some risk for each space journey. These risks change for each mission, depending on time, location, and the individual's health. Consideration of these influences led the Space Science Board and

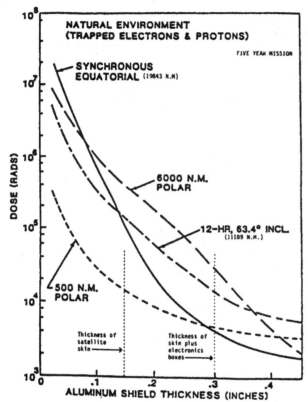

Figure 10.6   Accumulated radiation dosage over a five year mission due to Van Allen belt particles *(after Stauber et al., 1983)*.

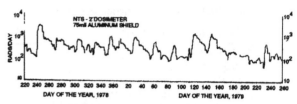

Figure 10.8   Fluctuation in daily dose for 12 hour (11,189 nm) orbit *(after Liemohn, 1984)*.

NASA officials to establish radiation dose guidelines for manned spaceflight exposure for a variety of mission durations. By combining selective mission planning with a proper degree of shielding, it was determined that man can perform effectively in space with minimal radiation sickness risk.

Several methods are available to mission planners to minimize radiation dose exposure. The "best" method is usually a compromise between mission profile, weight limitations, and crew health considerations.

Mission timing, when possible, can be useful. Mission timing may involve scheduling the construction of a space station for the least hazardous period of the solar cycle or minor changes to an existing mission such as termination of an EVA after very energetic solar flare activity. Unfortunately, this method can be very impractical based on need and commitment of funds and manpower.

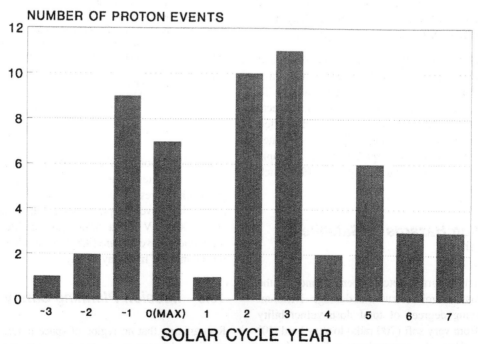

Figure 10.7   The average number of solar proton flares per year expected during a solar cycle. Year "0" refers to solar maximum; negative years are years before solar maximum and positive years refers to years after solar maximum. Notice the best chance of solar proton flares is in the second and third years after solar maximum *(after Hirman et al., 1988)*.

Orbital choice can be a viable means of reducing radiation exposure. In general, low equatorial orbits are the least hazardous, while geosynchronous orbits (GEO), especially during solar particle events, expose the crewmember to the greatest levels of ionizing radiation.

Vehicle shielding is another means of limiting the radiation hazard. External satellite skin thickness ranges from 0.8 mm of aluminimum (Al) to as high as several millimeters of Al. A large manned space structure would typically have about 5–8 grams/cm^2 Al shielding surrounding the habitat or work area. For most orbits, 5 gm/cm^2 Al represents a safe compromise between worker health and weight limitations. GEO receives the greatest shielding for various mission profiles. In GEO, for shielding greater than 1–2 gm/cm^2 Al, the Bremsstrahlung dose becomes the primary radiation source.

The need for greater protection during large solar particle events has resulted in several additional short-term protection schemes. One method involves orienting the surface of greatest shielding on the space vehicle toward the highest particle flux. Another method, the storm cellar approach, would allow heavy shielding in one region of the space vehicle, so that personnel could move into this "storm cellar" during hazardous solar events. Other less costly methods involve rearrangement of existing equipment in an attempt to protect the crew.

10.10 Environmental Effects of Space Systems

The deployment and operation of large structures in space will unavoidably be accompanied by the release of large amounts of expelled propellants into the Earth's space environment. The lift into low Earth orbit deposits large quantities of rocket exhaust products into the terrestrial atmosphere and ionosphere. Transfer to a higher orbit requires the release of rocket combustion products (or of energetic heavy ions if an electric propulsion engine is used) into the magnetosphere. Even when the spacecraft is in final orbit, both the spacecraft itself and its altitude control system are potential sources of released gases. These gas releases are of interest because they may interfere with the satellite system operation, and they may alter the natural environment.

Environment alteration is of concern for future space systems because these systems are expected to be much larger than systems in use at the present time. Several large systems have been proposed for deployment in the next two decades requiring structures of kilometer dimensions operating in very high orbits where the Earth's plasma environment is quite tenuous, and where the environment loading due to construction and operations by-products can be severe. Even neutral gases released at high altitudes may cause problems because these neutrals will eventually become ionized and these new ions may modify the plasma sheet, the magnetospheric current systems, and the magnetopause location.

One of the major impacts of these expellants is an alteration of the Van Allen radiation belt stability leading to a potential change of the radiation hazards to materials and personnel, and a modification of high energy particle precipitation events. If the addition of heavy ions increases the energy density of the radiation belts, then satellites passing through these areas will require additional shielding. The potential increase in auroral particle precipitation will modify the high-latitude ionosphere which, in turn, affects communication systems by increasing HF absorption and altering signal paths due to enhanced ionospheric refraction. The increased particle precipitation (and intensification of high-latitude current systems) will result in additional thermospheric heating. Such heating at high latitudes causes the thermosphere to expand upward and equatorward, resulting in increased drag for low-altitude satellites and reduced accuracy of satellite ephemeris predictions.

There is also growing evidence that solar variations affect weather and climate through subtle changes in geomagnetic activity. As shown above, a similar modification to the magnetosphere results from space system interactions, and there is the possibility that our extensive use of space may, in some way, affect the global climatic system. Unfortunately, our understanding of these magnetospheric weather links is so incomplete that it is impossible, at present, to estimate what effects space system modifications of the magnetosphere would have on the Earth's weather.

If there are harmful effects due the environmental modification by space systems, early identification of their impact will allow an opportunity to design alternate methods and techniques. In the final analysis, the economic and defense benefits of space systems need to be compared to the cost of environmental degradation, and also the cost of environmental effects by space systems needs to be compared to the possible problems resulting from the use of alternative Earth-based systems.

10.11 Summary

Spacecraft operations are ushering in a host of new problems. Electrostatic charging, and subsequent discharges, can befuddle operators of high-altitude satellites by generating false commands or damaging critical systems. Plasma injection and photoelectron emission are the primary sources of this problem. Plasma injection during a geomagnetic disturbance is also a major source of upper atmosphere density variations. Changes in solar emission can also alter the densities encountered by low orbit spacecraft. The resulting orbit changes can cause large tracking and positioning errors. While the above near-Earth orbit problems can confuse or damage spacecraft and even alter their orbits, space radiation can be lethal to astronauts. The wide variety of radiation sources make orbit planning particularly important. This need is reemphasized by the limitation of onboard shielding due to launch weight constraints. In all these areas, the observations are limited and their exact meanings are not yet clear.

10.12 References

Adams, J. H. 1981. Cosmic Ray Effects on Microelectronics Part 1,: The Near-Earth Particle Environment. NRL Report No. 4506, 25 August 1981.

Atwell, W. 1980. Dosimetry in the Manned Space Program, in *Solar Terrestrial Predictions Proceedings*, R. R. Donnelly, ed. NOAA/SEL, Boulder, Col.

Andrews, J. L., Schroeder, J. E., Gingerich, B. L., Kolasinski, W. A., Koga, R., and Diehl, S. E. 1982. Single Event Error Immune CMOS RAM, IEEE Trans. on Nuclear Sciences, vol. NS-29, No. 6, December 1982.

Baker, D. N., Belian, R. D., Higbie, P. R., Klebesadel, R. W., and Blake, J. B. 1986. Hostile Energetic Particle Radiation Environments in Earth's Outer Magnetosphere, AGARD Symposium, The Hague, Netherlands, 2–6 Jun 1986.

Banks, P. M., Williamson, P. R., and Raitt, W. J. 1983. Space Shuttle Glow Observations, *Geophys. Res. Lett.*, vol. 10, p. 118.

Blake, J. and Mandel, R. 1986. On Orbit Observations of Single Event Upset in Harris HM-6508 1K RAMSs, Aerospace Corp Report for Space Division, SD-TR-96-89, February 1987.

Bostrom, C. O., Fischer, C. L., Webb, P., Williams, D. J., Johnson, P. C., and Fabian, J. 1987. *Committee to Review Solar Flare Hazards to Man in Space*, USAF Scientific Advisory Board, Pentagon.

Browning, J. S., Koga, R., and Kolasinski, W. A. 1986. Single Event Upset Rate Estimates for a 16-K CMOS SRAM, Aerospace Corp Report for Space Division, SD-TR-86-99, September 1986.

Bueche, F. 1981. *Technical Physics*. Harper and Row, New York.

Burke, W. J. 1981. An Overview of Radiation Belt Dynamics, Proceedings of the Air Force Geophysics Laboratory Workshop on the Earth's Radiation Belts, October 1981.

Chapman, R. 1982. *The World in Space* (a Survey of Space Activities and Issues). Prentice Hall, Inc., N.J.

Chenette, D. L. 1986. Petersen Multipliers for Several SEU Environment Models, Aerospace Corp Report for Space Division, SD-TR-86-66, September 1986.

Cladis, J. B., Davidson, G. T., and Newkirk, L. L. 1977. *The Trapped Radiation Handbook*. Defense Nuclear Agency, General Electric Company, Santa Barbara, Calif.

Cunningham, S. 1984. Cosmic Rays, Single Event Upsets and Things that Go Bump in the Night. AAS publication No. AAS-84-05, February 1984.

Cunningham, S. 1985. Living with Things that Go Bump in the Night. AAS publication No. AAS 85-056, February 1985.

Garrett, H. B. 1980. Spacecraft Charging: A Review, in *Space Systems and Their Interactions With Earth's Space Environment*, H. B. Garrett and C. P. Pike, eds. American Institute of Aeronautics and Astronautics, New York.

Glasstone, S., and Dolan, P. 1977. *The Effects of Nuclear Weapons*. U.S. Government Printing Office, Washington, D.C.

Hess, W. N. 1968. *The Radiation Belt and Magnetosphere*. Blaisdell Pub. Co., Waltham, Mass.

Hirman, J. W., Heckman, G. R., Greer, M. S., and Smith, J. B. 1988. Solar and Geomagnetic Activity During Cycle 21 and Implications for Cycle 22, *EOS*, vol. 69, no. 42, p. 962.

Liemohn, H. 1984. Single Event Upset of Spacecraft Electronics, Proceedings of Spacecraft Anomalies Conference, NOAA, October 1984.

McNulty, P. J. 1981. Radiation Effects on Electronic Systems, Proceedings of the Air Force Geophysics Laboratory Workshop on the Earth's Radiation Belts, October 1981.

Marcos, F. A. 1991. Development and Validation of New Satellite Drag Models, AAS/AIAA Astrodynamics Specialist Conference, Durango, Col., Aug 19–22.

Mende, S. M., and Swenson, G. R. 1985. Vehicle Glow Measurements on the Space Shuttle, *Second Workshop on Spacecraft Glow*, NASA Conf. Pub.

Nightingale, R. W., Chiu, Y. T., Davidson, G. T., Francis, W. E., Rinaldi, M. A., Robinson, R. M., and Vondrak, R. R. 1986. A Space Radiation Test Model Study, Air Force Geophysics Lab. Tech. Report, 86-0064.

Nuclear Regulatory Commission. 1967. *Radiation Dose Levels for Apollo Crew Members*. File Memo FA 2-10-67, Publication 1487, Washington, D.C.

Petersen, E. L., Langworthy, J. B., and Diehl, S. E. 1983. Suggested Single Event Upset Figure of Merit. IEEE Trans. on Nuclear Sciences, vol. NS-30, no. 6, December 1983.

Rote, D. M. 1980. Environmental Effects of Space Systems: A Review, in *Space Systems and Their Interactions With Earth's Space Environment*, H. B. Garrett and C. P. Pike, eds. American Institute of Aeronautics and Astronautics, New York.

Rust, D. M. 1982. Solar Flares, Proton Showers, and the Space Shuttle, *Science*, vol. 216, p. 939.

Shea, M. A. 1984. Cosmic Rays and Heavy Ion Environments, Proceedings Spacecraft Anomalies Conference, NOAA.

Slanger, T. G. 1989. Vehicle Environment Interaction Workshop, *EOS*, vol. 70, no. 39, p. 859.

Smith, R. E., and West, G. S. 1983. Space and Planetary Environment Anteria Guidelines for Use in Space Vehicle Development, 1982 Revision (vol. 1), NASA TM-82478.

Stauber, M. C., Rossi, M. L., and Stassinopoulos, E. G. 1983. An Overview of Radiation Hazards in Earth Orbits, AAS Proceedings, Space Safety and Rescue.

Stevens, N. J. 1980. Review of Interactions of Large Space Structures with the Environment, in *Space Systems and Their Interactions With Earth's Space Environment*, H. B. Garrett and C. P. Pike, eds. American Institute of Aeronautics and Astronautics, New York.

Stevens, N. J., and Kirkpatrick, M. E. 1986. Spacecraft Environment Interaction Investigation, Air Force Geophysics Laboratory, Tech Report 86-0214.

Stoupel, E. 1980. Solar Terrestrial Prediction: Aspects for Preventative Medicine, in *Solar Terrestrial Predictions Proceedings*, R. F. Donnelly, ed. NOAA/ERL, Boulder, Col. p. G-29.

Townsend, R. E. 1982. *Source Book of the Solar-Geophysical Environment*. AF Global Weather Central, Offutt AFB, Neb.

Upton, A. C. 1991. Health Effects of Low-Level Ionizing Radiation, *Physics Today*, vol. 44, no. 8, p. 34.

10.13 Problems

10-1. Spacecraft discharge is *not* a significant problem for a spherical satellite with a homogeneous conducting surface. Why? What happens when dielectric materials (such as solar cells) appear on the surface of the otherwise conducting surface of the spacecraft?

10-2. Why are discharging events for geosynchronous satellites more common between 0400 to 0600 local time? Local time refers to what? Does the usual discharge local time change during geomagnetic activity? Why?

10-3. If atmospheric heating reduces the height of a circular orbit by 10%, determine the percentage change to the orbital period. Would a 1% change in orbital height make a significant change to the orbital period?

10-4. Why would a single period of heating have a permanent effect on an orbiting satellite? Does the heating have to be uniform or can it be localized? If so, give examples.

10-5. Why is the inner Van Allen belt most inflated (highest number density) during solar minimum? (Hint: see section 4.9.)

10-6. Why does the population of the outer Van Allen belts vary with the solar cycle? Does it make sense that the outer boundary of the electron belt is closer to Earth and contains higher energy particles during solar maximum?

10-7. Distinguish the differences between a rad, RBE, and rem. Which is the most effective tool for establishing biological damage from ionizing radiation? Why is ionizing radiation harmful?

10-8. What environmental factors should be considered when planning a space mission? Can the environmental dangers be minimized? How?

10-9. Can the deployment of space systems affect the structure of the space environment? If so, are these changes to the space environment potentially dangerous for future space operations?

Chapter 11

Space Weather Services
All material in this chapter is current.

11.0 Introduction

Over the past few years much has been written and said about the need and relevance of space weather forecasts and services for the U.S. commercial space user community. The nation's dependence on space-based systems continues to grow exponentially in order to meet the demands for improved communications and information delivery systems. Market forces in the user community demand an unprecedented level of accuracy in space weather forecasts. Unfortunately, the highly competitive nature of the space systems user community has prevented an open dialogue between forecast providers and the user community on what services are needed and the degree of accuracy required. The purpose of this chapter is to define these space weather nowcast and forecast requirements for the commercial space weather community. A nowcast is a short-range forecast usually on the order of 1 or 2 hours lead time.

11.1 The Commercial Space Weather Community

11.1.1 Community Overview

The commercial space weather community has two basic segments: consumers and service providers. As the name implies, consumers are the end users of space weather nowcasts and forecasts provided by the commercial space weather providers. The consumer segment is further subdivided into three components: system builders, system operators, and system users. The term *system* is used to denote any commercial system that is affected by space weather. For example, a system could be an on-orbit satellite, an electric power grid, long-haul communications, or global position satellite (GPS) geolocation systems. Space weather affects each of these systems in different ways, and therefore each system has its own unique need for space weather services. Even among users of the same system, needs vary by user due to differing business and operations needs. For example, User A's needs for a forecast of potential downlink disruptions to a communications satellite may be different than User B's needs for the same, or similar, satellites. Therefore, requirement analysis requires

more work than just picking a single member of a space consumer component and calling its needs representative of all members making up that component. Below are some general comments about the types of services required for each consumer component based upon our survey results

11.1.2 Space System Builders

In general, system builders are interested in worst-case engineering information. As might be expected, system builders want to understand the possible extremes of environmental conditions their systems will experience once fielded. The builders' intent is to minimize environmental sensitivities within cost and schedule constraints for fielding their systems. Many builders track the performance of their systems once fielded to establish any observed environmental sensitivities, both expected and unexpected. Oftentimes, this knowledge is folded back into their system redesigns, but is rarely shared with other system builders. This reluctance to exchange information is characteristic of most commercial enterprises which fear losing a competitive advantage within their respective communities.

11.1.3 Space System Operators

In general, system operators are interested in nowcasts and forecasts of space weather phenomena that degrade system performance. Typically, system operators buy or lease hardware systems from the builders and sell services to the system users. In some cases, builders own part, or all, of system operations companies. In other cases, a system builder may have acquired an operations company that uses a competitor's equipment. This smearing of lines between builders and operators is the result of a relatively small number of companies in each component.

11.1.4 Space System Users

System users make up the bulk of the space weather consumer segment. In particular, users want nowcasts and forecasts of system interruptions specific to those systems from which they buy operational services. For example, a television network may want to know when satellite communications links may be degraded or disrupted. The information comes from the satellite operations company with which it has contracted services. Generally, users desire such information to plan alternative courses of action if, and when, services are disrupted. The television network would want to plan ahead for alternative transmission links if the link it normally uses may be disrupted. Even though they are the largest space consumer component, the user community has the hardest time defining space weather nowcast/forecast requirements. A television network contracts services with a satellite operations company for satellite bandwidth availability at a reliability rate of, say, 99.9%. Such contracts have significant dollar penalties for the satellite operations company if it does not achieve the specified reliability rate. In this case, the television network's requirement is for satellite communications reliability. Space weather is just one of the many factors the satellite operations company has to consider when agreeing to such reliability rates. For the user, space weather requirements are qualitative rather than quantitative. We found that most members of the user community would be satisfied with the red/yellow/green type of forecasts where red is for potentially serious space weather impacts, yellow is for those times where there is a degree of uncertainty in forecasted conditions, and green is no problems anticipated. The exception is the insurance industry.

Companies who insure space systems want to know quantitative information about how space weather may (will) impact systems they are considering to insure. For systems already insured, the insurance companies are interested in establishing (after the fact) whether a problem was caused by space weather. Such determinations are often a key consideration in whether or not an insurance claim is paid. The insurance company needs highlight a systemic problem for anyone involved in providing space weather information to the user community. That is, without the cooperation of the systems operator and the builder, one cannot adequately assess whether space weather was or was not the causal agent of an anomaly. Determining the impact of space weather on any space system requires knowledge of how the system performed under similar space weather conditions in the past, for example. Although the user community is potentially the largest market area for any forecast service provider, it is the least accessible because it requires the full cooperation of the other two consumer components (operators and builders) who have little incentive to assist the forecast service provider.

11.1.5 Third-Party Nowcast/Forecast Service Providers

This is the newest segment of the commercial space weather community. Until recently, the government was the primary provider of space weather forecasts. Typically, these government nowcasts/forecasts were generic in nature and not tailored to specific space systems. For example, the government produced a generalized forecast on magnetospheric conditions, but did not attempt to say how such conditions might impact the performance of system A, B, or C. In fact, until recently space weather forecasting was not mature enough to be applied to specific space systems. With the advent of physics-based space weather models, such system-specific forecasts are now feasible albeit sometimes of poor forecasting accuracy. Since it is not the mission of the government to provide system-specific forecasts for commercial systems, the door was opened for what the government calls third-party vendors to offer tailored forecasting services to the commercial sector. The term *third-party* comes from the fact that these private forecasting companies are the bridges between government-provided information and the commercial consumer segments.

Besides immature forecasting technologies, there are legal considerations that discourage the establishment of a healthy third-party vendor community. For example, who is legally liable if a third-party vendor forecast is wrong, and a commercial space system is damaged? The long history of third-party commercial (tropospheric) weather forecast service providers gives some insight into the problem. Over the last three decades, no case law has been established by a court ruling between a third-party vendor and one of its customers over a faulty forecast. That is not to say that legal proceedings have not been initiated against third-party vendors. In fact, a number of legal suits were initiated and each case was settled out of court where the third-party vendor paid damages to its customers for the faulty forecast. In general, if the third-party vendor changes or adds to any portion of the government forecast, then the forecast liability rests with the third-party vendor. Adding content to the government forecast is at the legal risk of the third-party service provider. Such legal uncertainties are a definite deterrent for any third-party vendor.

11.2 Consumer Business Practices

11.2.1 Operations

In this section, we discuss how the space-system operators conduct daily operations. Understanding business practices is essential to knowing who establishes the operational requirements and who conducts operations. Sometimes the same group of individuals accomplishes both tasks, but generally separate groups conduct the two activities.

Visit any operations center for a communications, satellite, or electric power company and you will find a beehive of activities, which are mainly checklist driven. The demands of the job require operations center personnel to follow a set list of precise, time-synchronized activities. In the case where something unusual happens, operators follow carefully crafted contingency procedures. We specifically used the word *unusual* rather than unexpected because it is very rare that something truly unexpected happens. Through the years, personnel in the operations center have encountered almost every contingency and have developed a set of procedures to follow, should such an event occur again. Recurrent training on all contingency procedures is a regular part of all operations center personnel's scheduled activities. Any procedural changes require extensive additional training to ensure operations personnel will react correctly in all circumstances. In those cases where the unusual events do occur, the operations center personnel generally have just seconds to respond correctly; failure to follow correct procedures can mean the difference between recovering the system and damaging the system, sometimes permanently. Therefore, operations center personnel are generally skeptical of implementing any operational changes unless the benefits can be clearly demonstrated to them.

Generally, the needs (requirements) of the operations center are set at a decision level above that of the center personnel. This decision level is responsible for the cost/profit of the system(s) under control of the operations center. Operations requirements are driven principally by four factors: end-user service requirements, costs of doing business, available technology, and competition. End-user service requirements (system availability, reliability, bandwidth, etc.) are first and foremost. Generally, failure to deliver service at the specified level with the end-user can result in hefty revenue penalties for system operators. In order to maximize profits, system operators are always trying to minimize costs and still achieve the desired level of service. Two of the most common ways to minimize costs are to maximize system throughput (i.e., available bandwidth, system availability) and to keep operations personnel staffing at minimum levels. In order to meet end-user demands, system operators are always looking for new technology that can deliver more for less money. This, in turn, drives system builders to build more capability into a given system at the lowest possible price, which can increase a system's sensitivity to space environment. For example, building smaller, cheaper, and more capable satellites often requires builders to forego expensive (and throughput limited) radiation-hardened computer chips, and instead rely on more capable commercial grade chips. Although more computationally robust, these commercial grade chips are more easily disrupted by high-energy protons that crisscross the space environment. Finally, competition often forces system operators to raise their level of service to end-users. However, competition has a negative effect in that system operators become very secretive about revealing system sensitivities to space weather in fear that their competitors might get hold of this information and use it against them.

11.2.2 Consumer Attitudes about Space Weather

All three segments of the consumer community (builders, operators, and users) share a common myth that space weather either does not exist or doesn't pose a serious to space systems. Our interaction with the consumer community indicates the myth has four principal sources:

1. Belief in engineering invulnerability.
2. Ignorance about space weather hazards.
3. Competitive market place forces suppressing the facts.
4. Misconception that space weather is a fact of life for which there are no viable alternatives.

Sometimes multiple sources exist within the same company. Management may be convinced there is always an engineering solution, and competitive factors force companies to portray the image of invulnerability. In time, these denials become part of the company culture. Within the same company, the operations staff often are unaware the problems they are experiencing are related to space weather phenomena. In most cases, the "checklist mentality" trains the on-duty operations staff to deal with system anomalies without worrying about the underlying root cause.

The best way to debunk this myth is to demonstrate a real capability which can help space system consumers conduct their respective businesses more efficiently and effectively. The key is to show all consumer segments how they can save (or make more) money by using space weather forecasts. This generally requires a "fly before you buy" demonstration. That is, the space weather service provider (third party or government) needs to build a demonstration focused on the needs of a specific consumer to show how space weather services can be used in day to day operations.

11.2.3 Government Sector

In order to mitigate space weather impacts, the Department of Defense (DoD) and the National Oceanic and Atmospheric Administration (NOAA) maintain a worldwide network of ground-based and space-based sensors, which feed nowcast/forecast models. These government-provided observations and forecasts are available to third-party vendors who, in turn, tailor this information to the specific needs of the private sector. There are no official estimates of government savings associated with using existing forecasts and observations, although there is plenty of anecdotal evidence of savings. Since third-party providers are usually under nondisclosure agreements with their respective customers, there are no readily available published data on cost savings. The hope is that with reliable and accurate third-party tailored forecasts, most of the space weather incurred costs by the commercial sector today could be eliminated.

11.3   Commercial Sector Requirements

11.3.1   Methodology

Historically, workshops were the primary means the space weather research/science community adopted to attract space system engineers and/or operations center personnel into an open forum to discuss space weather requirements. Generally, such user community representatives cannot, or will not, state space weather product requirements. Engineers are interested in knowing the extremes of space weather hazards so they can design the next generation space system which is less susceptible to space weather hazards. Even if engineers are working on existing systems, they still are looking for engineering fixes to known sensitivities. Individuals from operations centers are interested in contingency procedures to assist them whenever system anomalies occur. They generally do not know, or sometimes do not care, if space weather is the source of the anomaly; they want to know what to do if, and when, anomalies occur. Therefore, as private consultants, we focused on those individuals who are responsible for the profit/loss for a given space system because they are responsible for delivering systems or services to their customers as efficiently and effectively as possible. These same individuals who would also make the decision on whether or not to actually buy space weather forecast services and integrate such services into day-to-day operations. These decision makers could also establish the needed accuracy of such services before they would consider starting a service.

As private consultants, we were able to establish legally binding prohibitions limiting our ability to associate space system deficiencies to any specific company. To those of us outside the space system industry, such levels of "secrecy" seem to border on paranoia. Nonetheless, secrecy is a fact of life in this industry and we had to find workable procedures to gather the necessary information. Due to the nature of our legal agreements, we had to guarantee anonymity for each company surveyed to gain its participation. Even with the safeguards we imposed, some companies still refused to discuss their requirements with us. Fortunately, enough companies from each consumer segment did talk to us so that we captured what we believe is a representative sample of space weather nowcast/forecast requirements.

After gathering the information from the consumers, the next step was to translate their requirements into meaningful statements of requirement that a space physicist or forecaster could understand. This turned out to be more difficult than we thought at the outset of our study. For example, satellite operations companies talk about requirements in terms of system availability, and not the fact that they need 2 MeV electrons forecast to a certain accuracy. Therefore, we had to look at system historical data (when made available) and deter-

mine how frequently systems experienced problems that were either attributed to space weather or fell into a user category of "problem cause unknown." We then ran an environmental model for those days when such problems occurred and assessed whether space weather really was the underlying cause. There were times when a company attributed a particular problem to space weather and we could not find a specific space weather cause using existing environmental models. That did not mean space weather was not the cause, but rather we could not identify the underlying root cause.

For those anomalies we could resolve, we established the system's sensitivity to specific space weather phenomena. We then made a projection of how space weather might affect the system in the upcoming solar cycle by using historical data from the last solar cycle. In the case of satellites, we examined 15 separate satellites over the last solar cycle. In this way we could determine how frequently space weather would impact a given system. We then calculated the expected system availability without space weather services. The next step was to fold-in anticipated improvements using space weather forecasts based upon "perfect" forecasts. We then backed off from perfect forecasts to determine the necessary skill score (using persistence as the reference forecast; see next section) to achieve the customer required system availability. For example, suppose a satellite operations company required 99% satellite availability to meet the needs of its customers. Through our retrospective study from the last solar cycle, we determined that space weather problems could adversely impact the system 2% of the time, giving us a system availability of 98%. Next, given perfect forecasts, we found one could achieve system availability of 99.5%. Then we determined what forecast accuracy we would need in order to achieve 99% system availability. Note that in some cases even a perfect forecast did nothing to improve system availability. Take, for example, the situation with a communications satellite when a company had no additional transponder capacity on other satellites to bridge the time when the space weather sensitive satellite was going to have problems. If we forecasted conditions that would disable a transponder, then even a perfect forecast would not improve system availability because the transponder would be off-line or degraded during the space weather event and there would be no alternative way to maintain services. It is apparent that our analysis required some judgment calls, but we developed a set of rules for our analysis and applied it equally to each system. Therefore, our final numbers might be a little low or high, but we expect the biases to be uniform across each consumer segment.

The average skill score required by the companies participating in this study was 0.8. At this skill score, these companies would fully integrate space weather forecast/nowcasts into their operations as a decision making tool. This means space weather information would automatically trigger contingency procedures whenever threatening conditions were forecasted to occur. A skill score takes into account not only the number of correct forecasts but also the number of missed forecasts and false alarms. A false alarm is when we forecast a problem will occur and then nothing happens. If a company still decided to use space weather forecasts with a lesser skill score, then contingency procedures would not be implemented automatically; instead someone would have to make a value judgment about what to do. A discussion of forecast value versus forecast accuracy is presented in Section 11.4.

11.3.2 Skill Score

As described in the previous section, we chose to use skill score rather than absolute accuracy to describe customer requirements. Skill score is the most common metric used by terrestrial weather forecasters to track quality improvements. Generally, forecasters track three quality metrics:

1. Accuracy: average degree of correspondence between a finite sample of matched pairs of forecast (f) and observations (x).
2. Association: Overall strength of the linear relationship between a finite sample of matched pairs of forecast (f) and observations (x).
3. Skill score (sometimes called relative accuracy): accuracy of matched pairs of forecasts relative to the accuracy of a reference forecast.

In mathematical terms accuracy is represented by the mean square error (MSE), denoted by

$$\text{Accuracy} = \text{MSE}(f, x) = \langle f_i - x_i \rangle^2$$

where the angle bracket is the statistical mean of the matched pairs f_i and x_i. Therefore, a perfect forecast would have a MSE of zero—the forecast and observations match perfectly. Usually, association is represented by the linear correlation coefficient of the matched pair sample (f_i - x_i). Skill or skill score is given by

$$SK_r = 1 - \frac{\text{MSE}(f, x)}{\text{MSE}(r, x)}$$

where r is the reference forecast compared to the same sample of observations (x). Typically, a reference forecast can be persistence (tomorrow is the same as today), or climatology, or some other forecast technique. A skill score of zero says the forecast technique f has no measurable improvement over the reference forecast method. Notice skill can be negative, meaning the forecast technique does worse than the reference method. A few examples will help illustrate the value of each metric, and why skill score is usually the metric used to track performance.

Suppose our goal is to forecast the geomagnetic index, Ap, 24 hours in advance. In addition, suppose the customer only wants to be notified when Ap exceeds 30 (minor storm threshold). We will use persistence as our reference forecast because this is the de facto method our customer uses. That is, the customer does not use any space weather forecasts, so by default the customer is assuming tomorrow will be like today. If we are to show this customer that space weather products are of value, then we must demonstrate our forecast model is better (i.e., has positive skill) relative to their existing technique. Sometimes, persistence is referred to as the "no-skill forecast." Suppose after a year, the reported forecast statistics are shown in Table 11.1.

In this example, the forecast is a simple yes (Ap will exceed 30) or no (Ap will be less than 30). Persistence had better "accuracy" and higher linear correlation (association) than the forecast method. But how much better is not readily apparent in either of these metrics. In this example, the forecast technique had considerably less skill than persistence and this lack of skill is obvious (negative skill score). The problem was a combination of false alarms (50) and missed forecasts (16). A false alarm is a forecast which did not verify (i.e., was not observed). Persistence, on the other hand, had more missed forecasts (18 compared to 16) but fewer false alarms (18 compared to 50).

Suppose we continue to refine the hypothetical data further and ask how well did the forecast do for major/server storms (Ap ≥ 50). Suppose further the results are shown in Table 11.2. In this case, our forecast method showed skill relative to persistence, specifically because of our reduced false alarm rate. Once again the relative improvement using the accuracy and association metrics is not so obvious.

Although in widespread use, skill scores can be misleading if the wrong reference forecast is used. Take, for example, the skill score for ring current electron fluxes during geomagnetic storms. Suppose the forecast technology, has an accuracy about twice the mean observation, or

$$\sqrt{\text{MSE}(f, x)} \cong 2x \quad .$$

That is, the forecast of ring current electron fluxes is off, on average, by a factor of 2. If we use persistence (p) as the reference forecast then we find

$$\sqrt{\text{MSE}(p, x)} \cong 10x \quad .$$

This gives us a skill score of 0.96 even though the accuracy's errors are huge. This excellent skill score is the result of an "unrepresentative" reference forecast. If instead we use the average values from the on-orbit satellite observations as the

Table 11.1 Statistics of forecasting when Ap will exceed 30. The upper two matrices are the truth tables for the forecast and persistence methods.

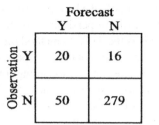

		Forecast			Persistence	
		Y	N		Y	N
Observation	Y	20	16	Y	18	18
	N	50	279	N	18	311

	Forecast	Persistence
Accuracy	0.425	0.314
Association	0.306	0.444
Skill	−0.832	

Table 11.2 Statistics of forecasting when Ap will exceed 50. The upper two matrices are the truth tables for the forecast and persistence methods.

		Forecast			Persistence	
		Y	N		Y	N
Observation	Y	10	1	Y	11	4
	N	3	351	N	8	342

	Forecast	Persistence
Accuracy	0.105	0.181
Association	0.827	0.663
Skill	0.663	

basis for a statistical climatology base, we find our skill score ranges between

$$0 < SK_c < .5 \quad .$$

The variance in the skill score depends on two factors: the time of the day for the forecast and the satellite type. Using the on-orbit climatology results in a skill score that is more representative of our real skill. Therefore, skill scores need to be interpreted in terms of the "representativeness" of the reference forecast.

11.4 Value versus Quality

Thus far, our analysis has focused on quality metrics because there is an underlying assumption that improving forecast quality increases the value of a forecast for the end-user. In fact, this is true to a point. Forecast value depends on multiple aspects of quality. Some lessons learned from the community at large can be summarized as follows:

- There is a quality threshold below which a forecast is of no value, but this lower threshold oftentimes is met well before capabilities meet the full stated requirement.
- There is an upper limit to the quality threshold for which increases in quality have no improvement in value.
- Both the upper and lower limits on quality evolve in time in concert with changes to end-user operations.

To demonstrate the first lesson learned, let's return to the example for major/severe storms (Ap $\geq$ 50) and suppose the end-user requirement is for a skill score of 0.8 (relative to persistence as the reference forecast). In the example (see Table 11.2), the forecast technique falls short of the desired requirement. Yet it is still fair to ask if the technique has value even though the quality does not meet the requirement. Suppose conversations with the end-user indicate they can live with false alarms, but missed forecasts are particularly troublesome. In this example, the forecast technique has only one missed forecast, compared to four for persistence (the customer's current de facto forecast). Therefore, the forecast technique would have value for this customer even though the skill score was only 0.663. As this example shows, value is very customer operations dependent, and value can only be delineated after considerable analysis on how customers conduct their business operations.

The second lesson learned seems, at first sight, to be counter intuitive. Suppose the end-user decision threshold is a skill score of 0.8. Providing a higher skill score gives everyone more confidence in the decision, but the decisions will be made at the 0.8 level. Therefore, spending more money to raise the skill score doesn't make a lot of sense when it won't influence the end-user's decision process.

The third lesson learned is obvious to anyone who works in operations. New technology, competition, and personnel training/availability are just a few of the factors that shape operational procedures. As we are all aware, these shaping factors change with time. For example, if an end-user loses all its senior operations staff, then the end-user might raise the required skill score in an effort to minimize both missed forecasts and false alarms until the staff gains more experience in handling unexpected system anomalies. In addition, as the system technology matures and experience grows, new environmental sensitivities may emerge. Competition also drives requirement changes because an end-user may need fewer false alarms to maintain higher system availability in order to offset lower prices to beat the competition.

11.5 Shortfalls

Our study also identified three major shortfalls that impact the ability of the private sector or government to develop services that can meet customer requirements. These shortfalls or deficiencies fall under the categories of science, observations, and funding.

11.5.1 Science

There are a number of promising high-resolution, gridded analysis and forecast models available today. These models are remarkable, considering the challenges facing the developers. First, space physics is a relatively young science with limited experimental data available to test theories and evolve the science. Second, space weather bridges a wide spectrum of space science disciplines which have not historically worked with each other. Accurate forecasts will depend on fully coupled solar-solar wind-magnetosphere-ionosphere-thermosphere models that can ingest a wide variety of observations often gathered in less than ideal locations. Finally, the third underlying cause is the failure to transition technology from the laboratory to operations. Just because something works fine in the laboratory doesn't mean it will work well in operations.

Transitioning technology to operations has to address three criteria: suitability, reliability, and usability. Suitability addresses whether the model is ready for operational use. In particular, can the model ingest current observations with their inherent errors, dropouts, and sometimes spotty delivery cadence? Reliability addresses the model's error management. For example, does the model amplify or smooth out errors inherent in the observations used to initialize the model? Do errors in one model output parameter spill over into other model output parameters? Finally, usability addresses the ease at which the model can be integrated into operations. How much training does it require? If the model crashes, how easy is it to recover? These are just a few of the examples relating to important issues that need to be addressed before a research-grade model can be transitioned to operations.

11.5.2 Observations

Regardless of the sophistication of space weather models, without timely and accurate observations to drive these models, the accuracy of these models will not improve. Presently, the operational community has access to limited real-time observations. In general, the government-run ground-based networks have limited geographic coverage and observation types. Space-based observations suffer from a limited number of satellites trying to cover a huge sampling volume. To say the space weather environment is undersampled does not do justice to the problem. In terms of a meteorological example, what we are doing in space weather today is equivalent to trying to forecast the sensible weather of the United States based only upon observations from Kansas City and St. Louis. Trying to forecast the troposphere's behavior based on such limited observations would be considered impossible, yet space weather service providers are faced with this situation every day. Much to the science community's credit, research instruments are now providing important observations to supplement the government's sensor networks. The problem with these research observations is they often have a limited lifetime, and if the sensor has problems, or data lines go down, for example, there is no mechanism in place to bring the sensor back online quickly. In addition, more effort needs to be focused on accessing international observations in near real-time to expand the observation baseline.

11.5.3 Funding

Funding shortfalls are the underlying problem that has had the greatest impact on both science and observations. As a discipline, space science has been historically underfunded. Through the efforts of the National Space Weather Program, interagency cooperation has helped bolster funding, or at least flatten the rate of decrease in government funding. The terrestrial weather community has done a much better job in keeping its funding line stable even in this era of shrinking government spending. In part, its success is due to two factors. First, everyone is touched by terrestrial weather, whereas space weather is at best an abstraction for most people. Second, terrestrial weather has a strong congressional lobby sponsored by the large community of third-party service providers. In space weather, third-party service providers are a silent constituency in Congress, and such groups are never rewarded with budgetary increases for their government counterparts.

11.6   Summary

This chapter has focused on articulating the space weather nowcast/forecast requirements for the U.S. commercial space community. The data collected were based on legally binding nondisclosure agreements with the companies visited. In addition, converting consumer requirements into statements meaningful to the space science community was a significant challenge. Such translation required extensive analyses including retrospective studies of possible space system disruptions by space weather. We also identified three shortfall areas (science, observations, and funding) that significantly limit our ability to meet the needs of the private sector.

11.7   References

Ehrendorfer, M., and Murphy, A. H. 1992. On the Relationship between the Quality and Value of Weather and Climate Forecasting Systems. *Idojaras*, 96, pp. 187-206.

Livezey, R. E. 1966. Some Useful Tools for Verification of Forecast and Models for Low Frequency Atmospheric Variability, in *The Evaluation of Space Weather Forecasts*, K. Doggett, ed. NOAA, Boulder, CO.

Murphy, A. H. 1996. Forecast Verification: A Diagnostic Approach, in *The Evaluation of Space Weather Forecasts*, K. Doggett, ed. NOAA, Boulder, CO.

Murphy, A. H., and Epstein, E. S. 1989. Skill Scores and Correlation Coefficients in Model Verification, *Mon. Wea. Rev.*, 117, pp. 572-581.

National Science Foundation Grant (Number: ATM-9610021)

Wilks, D. S. 1995. Forecast Verification, in *Statistical Methods in the Atmospheric Sciences*. San Diego, CA.

Index